Dr John Gribbin trained as an astrophysicist at the University of Cambridge before becoming a full-time science writer. He has worked for the science journal *Nature* and the magazine *New Scientist* (for which he is now Physics Consultant) and has contributed articles on scientific topics to *The Times*, the *Guardian* and the *Independent*. Gribbin has received awards for his writing in both Britain and the United States and is currently a Visiting Fellow in Astronomy at the University of Sussex. His books include *In Search of Schrödinger's Cat*, *In Search of the Big Bang*, *In the Beginning*, *In Search of the Edge of Time*, *In Search of the Double Helix*, *The Matter Myth* (with Paul Davies), *Stephen Hawking: A Life in Science* and *Einstein: A Life in Science* (with Michael White). Many of his books are published by Penguin. John Gribbin is also the author of several science fiction works, including *Innervisions*.

He is married with two sons and lives in East Sussex.

JOHN GRIBBIN

# IN SEARCH OF THE
# DOUBLE HELIX

## QUANTUM PHYSICS AND LIFE

PENGUIN BOOKS

PENGUIN BOOKS

Published by the Penguin Group
Penguin Books Ltd, 27 Wrights Lane, London W8 5TZ, England
Penguin Books USA Inc., 375 Hudson Street, New York, New York 10014, USA
Penguin Books Australia Ltd, Ringwood, Victoria, Australia
Penguin Books Canada Ltd, 10 Alcorn Avenue, Toronto, Ontario, Canada M4V 3B2
Penguin Books (NZ) Ltd, 182–190 Wairau Road, Auckland 10, New Zealand

Penguin Books Ltd, Registered Offices: Harmondsworth, Middlesex, England

First published by Wildwood House Ltd 1985
Published in Penguin Books 1995
1 3 5 7 9 10 8 6 4 2

Printed in England by Clays Ltd, St Ives plc

'If then we have under nature variability and a powerful agent always ready to act and select, why should we doubt that variations in any way useful to beings, under their excessively complex relations of life, would be preserved, accumulated, and inherited? Why, if man can by patience select variations most useful to himself, should nature fail in selecting variations useful, under changing conditions of life, to her living products? What limit can be put to this power, acting during long ages and rigidly scrutinising the whole constitution, structure, and habits of each creature – favouring the good and rejecting the bad? I can see no limit to this power, in slowly and beautifully adapting each form to the most complex relations of life.'

Charles Darwin,
*The Origin of Species*, 1859

'Only the fittest of the fittest shall survive.'

Bob Marley,
*Could You Be Loved*, 1980

If then we have under nature variability and a powerful agent always ready to act and select, why should we doubt that variations in any way useful to beings, under their excessively complex relations of life, would be preserved, accumulated, and inherited? Why, if man can by patience select variations most useful to himself, should nature fail in selecting variations useful, under changing conditions of life, to her living products? What limit can be put to this power, acting during long ages and rigidly scrutinising the whole constitution, structure, and habits of each creature,—favouring the good and rejecting the bad? I can see no limit to this power, in slowly and beautifully adapting each form to the most complex relations of life.

CHARLES DARWIN
The Origin of Species, 1859

. . . only the fittest of the fittest shall survive . . .

T.H. Huxley
'Evolution and Ethics', 1893

# CONTENTS

# INTRODUCTION

The idea for this book grew naturally out of my work on the story of quantum mechanics, *In Search of Schrödinger's Cat*. That book touched on the revolutionary impact quantum physics has had in the twentieth century on many areas of science, including the study of chemistry and especially our understanding of large molecules, the molecules of life. Without quantum mechanics, there would be no science of molecular biology at all. At about the same time, through my work with Jeremy Cherfas on two books about human evolution, *The Monkey Puzzle* and *The Redundant Male*, I developed my own understanding of the evolutionary debate since Darwin's time, and of the approach to genes and DNA from the opposite direction, outside inwards, so to speak, starting out from whole plants and animals and working down to the genetic material.

The story of Darwinian evolution has been told many times, not least as a result of the recent centenary; the story of DNA and molecular biology has been told almost as often. But I know of no account which does justice to the quantum roots of molecular biology, and few if any popular accounts tell the whole story of evolution, from Darwin to DNA, and beyond. My exploration in different contexts of the routes from quantum physics to molecular biology, and from Darwin to the genetic basis of evolution, suggested to me that there might be an audience for a book which tells the whole story and thereby provides, in an accessible way, the background to many of the scientific stories which make headlines today, from the creation 'debate' to genetic engineering. I hope that this book fits the bill.

I have not tried here to argue the case for the reality of evolution itself, preferring to let the facts speak for themselves. But since there are those who still try to argue the case against evolution it may be of interest, even in the middle of the 1980s, to notice just how well the whole story hangs together. Quantum physics and molecular biology provide an explanation of the mechanisms of evolution, which Darwin himself knew must exist but which remained uninterpreted in his day. They explain exactly how genetic information is passed on from parent to offspring, and just why occasional mistakes in copying this information occur, so that the offspring do not always perfectly resemble their parents. And this accurate, but not quite 100 per cent accurate, copying of the genetic message is the very stuff of Darwinian evolution, as we shall shortly see.

Before I begin my tale, though, I should acknowledge the many friends and colleagues without whom this book – something of a departure for an author whose own scientific background is in physics and astronomy – could never have been written. My wife, herself trained in the biological sciences, provided help even more valuable than her contributions to my previous books, while Steve Lee and his colleagues at the University of Sussex Library have recently developed a computer-based catalogue which greatly simplifies the search of the scientific literature essential in preparing a book like this. I don't know how I ever wrote a book in the past without it – just as I now find it impossible to comprehend how I ever wrote books without the aid of a word-processing computer, or a photocopier. Gail Vines, of *New Scientist*, nobly read through the whole text in draft, and saved me from the embarrassment of perpetrating too many biological howlers, while Lionel Milgrom did a similar job on the chemical chapters. And although he made no direct contribution to the present book, I am indebted to my biological mentor, Jeremy Cherfas, for opening up a new world to me. Finally, I will always be grateful to my younger son, Ben, for a comment he made when I was halfway through the first draft, with no end to the project in sight, and assailed by the usual doubts authors have about whether they are wasting their time. He picked up a stack of printout, sat quietly reading it for more than an hour, then turned to his mother with these words: 'I like this book. It's really interesting. Even if you don't understand all the words, it's still like a story book.' It is things of such nature that keep us writers going; I hope you enjoy the book as much as he did.

John Gribbin
June 1984

# PART ONE

# DARWIN

'Why, if man can by patience select variations most useful to himself, should nature fail in selecting variations useful, under changing conditions of life, to her living products?'

Charles Darwin
*The Origin of Species*, 1859

# CHAPTER ONE

# DARWIN UP TO DATE

Charles Darwin is widely regarded as the father of the greatest intellectual revolution of all time. His theory of evolution by natural selection, published in *The Origin of Species* in 1859, challenged the widely held views not just of science but of society as a whole. Indeed, Ernst Mayr, Professor of Zoology at Harvard University, said in his contribution to the Darwin Centenary Conference in 1982 that 'there could not be any truly objective and uncommitted science until science and theology had been cleanly and completely divorced from each other', and that the publication of the *Origin* was the greatest single influence in bringing about this divorce.* The extent of that revolution in the hundred years since Darwin's death is clear from the fact that I can begin this book with such a statement, without explaining to you who Darwin was, or what the theory of evolution is. I am confident that every reader will have some idea of both the man and his work, in a way that I could not be confident that every reader of a book on cosmology, say, would have even the most hazy grasp of Einstein's general theory of relativity. How far your image of Darwin's work matches up with mine remains to be

---

* See Mayr's contribution to *Evolution from Molecules to Men*, edited by D. S. Bendall. Full details of works cited are given in the Bibliography.

seen; we can agree from the outset, however, that by explaining the origin of all species – including man – in terms of natural processes, operating throughout history exactly as they do today, Darwin most certainly did divorce biological science, at least, from religion.

But was this revolution uniquely 'Darwinian', or was it an inevitable product of its time? Was science, in fact, ready to break with religion in any case, with Darwin doing no more than to hurry the process along? It is no slight on Darwin's undoubted genius, and especially his ability to marshal evidence from across a wide variety of scientific disciplines to make a cogent case, to point out that by the middle of the nineteenth century the time was very much ripe for the idea of natural selection – as proven by the fact that Darwin's contemporary, Alfred Russel Wallace, independently came to the same idea, prompted by much the same observations as those Darwin had made.

In many ways, Darwin was a timid revolutionary. Although his theory was essentially complete by 1842, six years after he returned home from his famous voyage round the world on the *Beagle*, he kept it largely to himself until 1858, when he received a letter from Wallace, in Borneo, outlining essentially the same theory. It was fear of being pre-empted by Wallace that forced Darwin to publish. And even then, a semi-invalid, he kept his head down at his writing while Thomas Henry Huxley, a brilliant young anatomist (who commented, on first learning of Darwin's theory, 'how extremely stupid not to have thought of that'),* fought the case for evolution in public debates, so fiercely that he became known as 'Darwin's bulldog'. Without Darwin, Wallace would still have published the theory of natural selection; without Wallace or Darwin, Huxley or one of his contemporaries would indeed have thought of it, sooner rather than later. The theory of evolution by natural selection was an inevitable

---

* See Huxley's contribution to *Life and Letters of Charles Darwin*, edited by F. Darwin, volume 2.

product of the improved understanding about the geological history of the Earth, its age, and the nature of fossil remains, that had been growing up throughout the eighteenth and early nineteenth centuries.

# The gift of time

The great gift that geology bestowed upon Darwin's generation of scientific thinkers was time. At the end of the eighteenth century, Christian theologians still taught that the age of the Earth could not be more than about 6 000 years, and that the Creation had occurred in 4004 BC. This conviction came from a literal interpretation of the Bible, counting back the successive generations to Adam. With such a short span of time to play with, those who thought about such things had to imagine either that the Earth was unchanging, or that changes were brought about by sudden, dramatic catastrophes, such as the Biblical flood. When evidence began to come in that the Earth was much more that 6 000 years old, and that its surface had been changed dramatically during its long history, it was, perhaps, natural that catastrophism should initially gain the upper hand as not one Biblical flood but many different catastrophes were invoked to explain the new geological findings.

Those findings were, indeed, dramatic. The English surveyor William Smith, whose work in the late eighteenth century took him down mines and along the cuts of new canals, was the first person to make a systematic study of fossils. He kept notes of the different layers of rock that he came across (the strata), and of the fossils they contained, eventually establishing that each stratum, wherever found across the country, contained a characteristic group of fossils, the remains of characteristic species of living organisms. The implication, picked up by many geologists and thinkers of the time, was that the Earth must be very old, that successive strata had been laid down one by one as time went by, and that different forms of life had appeared, lived for a time on the face of the Earth and then been replaced by others, many times throughout this long span of history.

In the early nineteenth century these ideas were incorporated into a theory of catastrophism by the French scientist Georges Cuvier. He saw the divisions between strata and the differences between the lifeforms they contained as signs of sudden, lethal changes in the environment on Earth – changes in sea level that brought flooding, upheavals producing new mountain ranges, or rapid changes in climate. After each upheaval, new life forms had to be specially created by God, only to be destroyed, in due course, by the next catastrophe. This was the established world view until well into the nineteenth century. But there was an alternative view, one which had a great influence on Darwin.

## Lyell's influence on Darwin

James Hutton was a gentleman scientist of the old school. Born in 1726, he made enough money out of farming and the invention of a process for manufacturing the chemical sal ammoniac that by 1768 he was able to devote himself almost entirely to his scientific meetings, and field trips to study different kinds of rock formations. All of this led him to conclude that the effects of natural processes that operate on the Earth today – erosion by running water, the action of tides, volcanic outbursts and so on – were sufficient to explain the changes that had taken place in the surface of the Earth since it first formed, provided that these natural processes had a sufficient span of time to work with. This is the essence of the idea of uniformitarianism, the very antithesis of catastrophism. To Hutton, no special acts of violence were needed to explain the geological structures we see today, just everyday processes operating for an extremely long time.

Hutton's ideas were presented to the Royal Society of Edinburgh in 1785, and a second paper was presented three years later. His principle assumed an enormously long span of time for the history of the Earth, much greater than that required by the catastrophists, and it was hardly acknowledged as a serious possibility by his contemporaries. It wasn't until 1793 that the establishment took sufficient notice of his views to attack them with much vigour. Hutton's response was a two-volume *Theory of the Earth*, published in 1795; his death in 1797 prevented completion of a third volume, but a succinct account of the theory of uniformitarianism was published by his friend John Playfair in 1802. The world of science remained unconvinced: catastrophism remained the established wisdom of the day, and if anything *gained* strength through Cuvier's work, but at least the idea of uniformitarianism had been aired, and had its proponents. That was the background to the geological work of Charles Lyell, which was to be the single most direct influence on Darwin's scientific thought during and immediately after the voyage of the *Beagle*.

Lyell, another Scot, was born in the year Hutton died. And, like Hutton, his scientific career did not follow the conventional academic route that we take for granted today. He went up to Oxford University at the age of 19, graduated in 1819, and set off to London to study law. But although he was admitted to the Bar in 1825, both his studies and his practice of law suffered from his enthusiastic, but strictly speaking amateur, interest in geology, a passion he was able to indulge thanks to the financial support provided by his father. He developed a plan for a book which would argue that all geological phenomena could be explained by natural processes, without invoking the supernatural, and in 1828 he set off on an expedition round Europe to gather material in support of this contention. From France and Italy, and especially the volcanic region around Mount Etna in Sicily, he found ample confirmation of the power of natural forces. Returning to London in February 1829, he began work on his great book, *Principles of Geology*, and saw the first volume published in 1830. By July 1830 Lyell was off on his travels again, this time to the geologically interesting region of the Pyrenees and south into Spain. The second volume of *Principles* duly appeared in 1831, and the third, and final, volume in 1833.

Lyell's work built from that of Hutton, but went further. Even so, it did not sweep catastrophism away overnight. It took decades for uniformitarianism to become established as the guiding principle of geology. But Lyell's book did cause immediate debate, and made many immediate converts. One of them was Charles Darwin, just 12 years younger than Lyell, who took the first volume of *Principles* with him when he set off on the *Beagle* on 27 December 1831. The second volume caught up with him later in the voyage: volume three awaited his return in 1836, when he was still only 27. Darwin gained two things from Lyell – a long span of time during which natural processes could operate, and the idea that nothing more than the natural processes we see today, operating over that long span of time, is required to explain great changes. Equally, Lyell's scientific approach to natural problems, his way of ordering the material and presenting an argument, also made a deep impression on Darwin, who was later to write:

> 'I always feel as if my books came half out of Lyell's brain, and that I have never acknowledged this sufficiently . . . I have always thought that the great merit of the Principles was that it altered the whole tone of one's mind.'*

What, then, was Darwin's own background, and the background to the idea of evolution, that placed this young man of 22 in the ideal situation – as naturalist on a voyage around the world, armed with Lyell's newly published *Principles* – to develop the idea of evolution by natural selection?

## The image and the man

In popular mythology, the young Charles Darwin is often presented as something of a caricature. He came from a wealthy family. On his mother's side from the

* Letter cited by Jonathan Howard in *Darwin*.

Wedgwoods who established the famous pottery, while his paternal grandfather, Erasmus Darwin, was, among many other things, also a famous natural philosopher of his day, who speculated about the nature of biological evolution. The popular image, handed down in many biographies and in TV popularisations, is of a young man more interested in huntin', shootin', and fishin' than in the serious things of life, an early nineteenth-century 'sportsman' typical of the younger sons of newly-rich families of the time. But this scarcely does credit to the man. Recent biographers have pointed out that the image owes much to Darwin's own recollections, much later in life, and that these reminiscences, though perhaps indicating modesty, are not always completely reliable. In fact, all the evidence is that, although he did enjoy his sporting activities, from an early age Darwin took a serious interest in the world about him, that he was a committed man of science and well aware of the important new developments in geology.*

Certainly Darwin failed in his attempt to become a doctor, as his father had wished, flunking out from Edinburgh University after seeing operations performed without anaesthetic. But while in Edinburgh he was already studying geology, and when he went to Cambridge, ostensibly to work towards a career in the Church, he had ample opportunity to develop this interest and his fascination with the living world. He became a close friend of the Professor of Botany, John Stevens Henslow. Even in the early nineteenth century, Cambridge professors were not in the habit of admitting dilettante playboys to their inner circle, and Darwin's close relationship with Henslow alone speaks volumes for his dedicated interest in science by that time. It was through Henslow's influence that, in the nick of time before being required to take Holy Orders, Darwin was offered the post of (unpaid) naturalist and Captain's companion on the *Beagle*, not by luck or as some lark to pass the time, but because the Cambridge Professor of Botany

---

* See Peter Brent's excellent biography, *Charles Darwin*.

thought that he would be the ideal man for the job. The Professor's confidence was fully justified by events.

Of course, Darwin *was* lucky to have been born into a wealthy family, where his father was able to support him through two universities and, with a little persuasion from young Charles' maternal uncle, Josiah Wedgwood II, to allow him to depart on a long sea voyage instead of begining a 'proper' career at the age of 22. Few people then, or since, have had the advantage of such a start in life; but then, few of those who *have* had such opportunities have made half the use of them that Darwin did. Details of Darwin's life can be found in Peter Brent's biography; his own description of events on the voyage of the *Beagle* can be read in a collection conveniently packaged by Christopher Ralling. What I am concerned with now is the scientific background that paved the way for Darwin's own great work, a background which came largely from Lyell.

Geology and biology were already inextricably linked before Lyell began his work. Characteristic fossil remains help to date geological strata, and the changing conditions revealed by differences between the layers of rock provide a clue to why the fossils change. Lyell argued that any species must become extinct sooner or later, because the changing conditions on Earth will destroy the habitat – the ecological niche – to which it is adapted. He argued that the nature of each and every species is tailored to a particular set of environmental conditions that it needs to flourish, and that when that set of conditions disappears so does the life form. But he rather begged the question of where new life forms come from to fit the new ecological niches created when conditions change. He specifically argued against the possibility that one species might be transformed into another, by 'transmutation', and simply said that new species are created to fit the new conditions. There was still a role for God in Lyell's world, not just a once and for all creation, but almost day to day tinkering with life on Earth to ensure the presence of species to fit every ecological niche.

This was the background to Darwin's own thinking, his point of departure. Darwin's ideas on evolution progressed from Lyell's in the light of what he saw on his voyage around the world. Darwin was a great observer, with an eye for detail and relationships. He saw that species in younger lands, especially the birds of the Galapagos Islands, resembled very closely the species on nearby older lands, and that the species on neighbouring islands resembled each other even more closely than they did those on the mainland. Although each of those island species *is*, indeed, exquisitely adapted to its own ecological niche, he saw that the *origin* of those species could best be explained in terms of descent from a common ancestor, a migrant species from the mainland, with the common descent explaining their similarities and the exquisite 'tailoring' to fit different niches explaining their differences. Adaptation and ancestry together can explain far more than adaptation alone; but what was the *mechanism* through which descendants were gradually modified into forms different from those of their ancestors?

# Darwin and Malthus

This was the central puzzle in Darwin's mind when he arrived home in 1836, at the end of the five–year–long voyage of the *Beagle*. One of the clearest, most authoritative reconstructions of how he developed his ideas on natural selection over the next six years has been carried out by M. J. S. Hodge, of the University of Leeds, who summarized his findings in the centenary volume edited by D. S. Bendall. According to this reconstruction, the key developments in Darwin's own thinking occurred over a period of about six months, beginning on 28 September 1838, when he first read the now famous *Essay on the Principle of Population*, by Thomas Malthus. This work is widely misunderstood and misrepresented these days, notoriously by the doom-mongers of the 'limits to growth school', who see the world imminently in danger of going to hell in a handbasket because of the existence of

'Malthusian limits'. So it is worth stepping back from Darwin's own story, briefly, to check what Thomas Malthus really said.

Malthus was born in 1766, and published the first edition of his *Essay* in 1798, at that time anonymously. Being a trained mathematician, he was intrigued by the way in which human population, and populations of other species, would increase if allowed to do so unchecked. He realised that the nature of this increase is geometric – that is, it doubles at regular intervals instead of increasing in a steady, or linear, way. Such an increase has explosive potential if unchecked, as can easily be seen by imagining a world in which no checks and balances restrained the increasing population of rats, or cockroaches, or rabbits. Although human beings reproduce more slowly than these species, Malthus pointed out that in the new lands of America at that time, population was actually doubling rather more quickly than once every twenty-five years. All this requires, of course, is that each couple should, before the age of 25, produce enough children so that four of them survive to have their own children. And yet, quite clearly, human population has not been doubling every 25 years throughout history. Even the slowest breeding animals of all, the elephants, could, if unchecked, produce 19 million offspring in only 750 years; yet, until 'civilised' man came along to upset the existing balance, the population was more or less constant. On average, each pair of elephants from 750 years before left only one pair of descendants every 750 years, or any other number of years, later. Why?

Malthus realised that all populations are held in check by counterbalancing forces – the attacks of predators or, most crucially, the supply of food. Population expands to consume the resources available, or as he put it:

'The natural tendency to increase is everywhere so great that it will generally be easy to account for the height at which the population is found in any country. The more difficult, as well as the more interesting, part of the inquiry is to trace the immediate causes which stop its further progress. . . . What then becomes of this mighty power . . . what are the kinds of restraint, and the forms of premature death, which keep the population down to the means of subsistence.'*

Since human – or elephant – population has not increased throughout history at the appropriate geometric rate, something is holding it in check. Malthus identified that something in terms of famine, pestilence, war and disease. His gloomy prognostication was that by reducing the effect of these natural checks in, as he saw it, a misguided attempt to improve the lot of humanity, we would only be storing up trouble for ourselves. Only 'vice' (which, for him, included contraception), 'misery' or 'self-restraint' could check the headlong growth of human population, a growth which farmers would be unable to keep up with in terms of food supply.

That debate continues. It's worth noting, in passing, that since the 1940s, during the most explosive period of growth in human population, growth in food production has, in fact, run at a slightly faster rate.† And, of course, contraception is no longer quite so obviously a 'vice' as it was to Malthus. But that is as far as our detour from Darwin's thinking takes us for now.

# Three keys to evolution

The key to what Malthus' *Essay* meant to Darwin is contained in the term 'premature death'. Malthus pointed out that the *majority* of all individuals, in the natural state, do not survive long enough after they have been born to breed in their own turn. Darwin wondered why some individuals, the minority, should survive and reproduce, while others did not. And he saw clearly that the most successful individuals would be the ones that were best adapted to their particular ecological niche. Putting it the other way round, the least well adapted would be losers in the struggle for

---

* See Antony Flew, *Malthus*.
† See my book *Future Worlds*.

what Malthus pointed out were limited resources. To use
the term now so familiar to us, the individuals best fitted for
their niches would survive. The superfecundity of life,
coupled with rigorous culling, left only those individuals
most suited for the environment.

On the day that he first read Malthus' *Essay*, Darwin
wrote in his own 'Notebook on Transmutation of Species'
the following:

> 'On an average every species must have same number killed year
> with year by hawks, by cold, &c. – even one species of hawk
> decreasing in number must affect instantaneously all the rest. The
> final cause of all this wedging must be to sort out proper structure
> . . . there is a force like a hundred thousand wedges trying to force
> every kind of adapted structure into the gaps in the economy of
> nature, or rather forming gaps by thrusting out weaker ones.'*

The struggle for survival operates, however, not
between *species* but, as Darwin quickly appreciated,
between *individuals*, different members of the *same*
species.

The next stage in the development of Darwin's ideas
came around the end of November 1838, when he first
began to make the analogy between this process occurring
in nature and the process of selection whereby mankind
has adapted other species to his purposes – dogs, horses
and so on. By choosing in each generation the dogs with the
longest legs, or the biggest horses, and breeding only from
those, man had been able to 'create' the greyhound and the
carthorse from different ancestral forms. Nature could do
the same trick, by 'choosing' the varieties best fitted for a
particular niche and leaving the others to die – hence
'natural selection', by analogy with man's artificial selection of species. There was just one more step needed to
complete the basic theory.

---

* See 'Darwin's Notebooks on Transmutation of Species', *Bulletin of
the British Museum (Natural History), Historical Series*, volume 2, 1960,
and volume 3, 1967. The biography of Charles Darwin, by Gavin de
Beer, discusses the development of Darwin's ideas in detail and is more
accessible.

Natural selection can only distinguish which individuals are more fit, in this sense, if there is a variety to choose from. Early in 1839, Darwin reasoned that natural selection did not have to 'know' what the objective of evolution was, but that it could 'work' with variation that arises accidentally. If it just happens that a bird is born with a slightly longer beak, and if that beak helps it to find food, then there is every chance that it will survive to pass on the characteristic long beak to its own offspring. Birds with shorter beaks will be progressively squeezed out of that particular niche, because the longer-beaked variety will eat all the food.

These are the keys to Darwin's theory of evolution by natural selection. It operates on individuals, allowing only those best fitted to a particular environment to survive; it involves inherited characteristics, that are passed from one generation to the next; but that process of inheritance is imperfect, so that nature has a variety of individuals in each generation upon which to practise the selection process. Darwin had reached this point in the argument by March 1839, and had put it in writing; in 1842, when he was 33, he wrote out a much more complete essay, his *Sketch*, which was published by his son Francis Darwin in 1909. But from 1842 until he was jolted into action by the arrival of Wallace's letter in 1858, he kept the idea to himself, sharing it only with a few close friends and trusted scientific colleagues.

# Darwin and Wallace

Wallace, the co-founder of the idea of evolution by natural selection, had been born in 1823, and by the 1840s, when Darwin was completing his own private account of evolution, he was a schoolteacher with a passion for botany and collecting plants. This interest soon extended to insects as well, and in 1848 Wallace went on a scientific expedition to the Amazon, where his observations of the profusion of life in the tropics led him to support the idea of evolution, though as yet with no better ideas than Lyell on how species came to be adapted to their niches. Pursuing this interest, in 1854 Wallace set off on a long visit to the Malay Archipelago, where he stayed for eight years. The variety of tropical life on these islands, and the differences from one island to the next, led him along much the same path of reasoning that Darwin's travels in the *Beagle* had led him, a quarter of a century before, with the difference that Wallace had already read Malthus' *Essay* before he set off on his expedition. It was while recovering from a severe attack of malaria, in February 1858, that he remembered this book and put it in the context of evolution. His biographers tell how the idea of survival of the fittest came to him in a flash, and how the essay which was to have such an impact on Darwin was written out over the next two evenings and sent to Darwin by the next mail.

Darwin's reluctance to make his ideas public was, of course, prompted by the certain knowledge of the wrath it would bring down upon his head from the Church and those who believed in a literal interpretation of the Bible. More particularly, he was probably concerned to avoid, as far as possible, offending the susceptibilities of his wife, who still retained the religious convictions which Darwin himself had long since lost. But he could no longer sit back and keep his views to a small circle of friends if Wallace was about to publish. Concerned to know how best to deal with this puzzle, he turned to Lyell (by now Sir Charles Lyell) and the eminent botanist Sir Joseph Hooker. At their advice, a combination of Wallace's essay and a summary of Darwin's ideas on evolution was presented as a joint paper to the Linnean Society in London on 1 July 1858.

This is occasionally represented by modern writers as a mean trick to play on Wallace, who, if he had happened to send his essay to someone else rather than Darwin, might have published first, alone, and gained all the credit. But it can be argued, with equal force, that Darwin did not choose the easy option of sitting back to let Wallace bear the brunt of the inevitable attacks on the idea, and stood up to be counted alongside his colleague. Certainly, Wallace has his place in the scientific halls of fame today, a place he rightly earned through his own perception of evolution at work. In his section of the joint paper of 1858, he made all the points outlined above in connection with Darwin's work, and concluded that 'those [individuals] that prolong their existence can only be the most perfect in health and vigour . . . the weakest and least perfectly organized must always succumb'.*

In spite of such clear statements, the Linnean Society was not immediately impressed. It happened that this particular meeting, the first at which Lyell and Hooker could get the paper read, was a special one called to elect a new Vice-President, so that the members' minds were not fully on the scientific part of the meeting. Also, a rather large number of papers were read at the meeting – six in all – so that the members may have been rather overwhelmed by the mass of information they received. Even so, with hindsight it is rather remarkable that eleven months later, on 24 May 1859, the President of the Linnean Society summarised the twelve months just passed with the comment 'the year . . . has not, indeed, been marked by any of those striking discoveries which at once revolutionise, so to speak, the department of science on which they bear'.† Six months later to the day, on 24 November 1859, the first edition of the *Origin* appeared and promptly sold out. By 1872, it had been through six editions; long before then, the revolutionary impact of what Darwin and Wallace presented to the Linnean Society in 1858 was clear far beyond scientific circles.

# Science and belief

Without going into details of the nineteenth-century debate about evolution, there is one crucial point which illustrates the scientific method at work and which should

* Reprints of the early papers can be found in H. L. McKinney's *Lamarck to Darwin*.
† *Journal of the Proceedings of the Linnean Society, Zoology*, volume IV, page viii, 1860.

be stressed. Opponents of the idea, then as now, were generally people who *believed*, as an act of faith, in the Biblical story of creation. It happens that this story was already very much under attack long before Darwin's day. It is one thing to imagine a Creator who did the job once and for all and left things to proceed in line with the rules He had established, but quite another, as Darwin's immediate predecessors had realised, to envisage a Creator who kept making mistakes, wiping out whole families of species in repeated catastrophes, then creating new ones to slot in to the new ecological niches. This is more characteristic of the shoddy builder who keeps patching things up, rather than of the great architect working to some grand design. For those who do subscribe to the religious view of creation, it is quite possible to reconcile this with the idea of evolution by taking the Biblical story as allegory and imagining the Creator establishing the whole Universe, with all its physical laws, and then leaving evolution, both physical and living, to take its course. That truly does smack of the Great Architect. But that is not the crucial point about Darwin's scientific approach to problems.

Darwin pointed out many times, including in his letters later collected by Francis Darwin, that he did not *believe* anything. Like all good scientists, he created working hypotheses to explain his observations of the natural world, and then looked to see how well the rest of the world fitted in with those hypotheses. He regarded the theory of evolution by natural selection as a good working hypothesis, because it could explain so many phenomena that were otherwise inexplicable, except by the actions of a somewhat capricious Deity constantly tinkering with nature. The distinction sounds subtle, but is crucial. Ask devout Christians whether they *believe* that Christ died and rose again, and they will say that of course they do. Ask them for evidence, and they will be baffled by the question. It is not a matter of evidence, but of *belief*; asking for evidence indicates doubt, and with doubt there is no faith. But science is, or should be, all about doubt. Ask a scientist whether he *believes* in evolution, or that the Earth is round,

and when pressed, if he is a good scientist, he will admit
that these are good working hypotheses, but that new
evidence may yet emerge which requires them to be
replaced by better hypotheses. Science and religion speak
different languages, and that is why the debate between
'creationists' and 'evolutionists' was, and is, ultimately
sterile.

The point is especially important because one of
Darwin's hypotheses was very much incorrect. The fact
that it was later tested and found wanting, and replaced by
a better hypothesis, indicates the strength, not the weak-
ness, of the scientific method. Unswerving *belief* in the
Gospel according to Darwin would have held back progress towards a better understanding of nature.

# Variation and evolution

This key development concerned the origin of the variation
between individuals which is the raw material upon which
natural selection can act. Without variation, there would
be no differences to select from, and although many individuals would still die, because of the Malthusian pressures
upon them, their deaths would be random, with no influence on the overall characteristics of the species in later
generations. Variation and heritability are as important to
evolution as natural selection itself.

Darwin suggested that evolution operated in two ways.
First, it kept a species exquisitely tailored to its own ecological niche, by singling out any tiny advantages which one
individual possessed – the longer beak, in the example I
used above. But this was the lesser aspect of natural selection, as indicated by Darwin's choice of title, the *Origin of
Species*. Darwin explained that when groups of individuals,
initially members of the same species, became separated
from one another then each group would continue to
evolve.

In the case of the birds of the Galapagos Islands, a few
original settlers from the mainland were presumably members of the same species (the ones chosen by Darwin for

particular study were different varieties of finch). Different individuals settled on different islands, where the slightly different environments provided a selection pressure for different physical features. On one island, perhaps, the longer beak was an advantage, and so steadily evolved; on another, a short, thick beak might have been advantageous for cracking open the kind of seeds found there, and over many generations the descendants of the original settlers would evolve in that way. In due course, the two families of birds, descended from originally identical ancestors, would become quite distinct, with different feeding habits and different physical appearances. This process, argued Darwin, operating over a long enough timescale could explain not only the differences between closely related types of finch, but also between lion and tiger, man and ape, man and tiger, tiger and finch, and all other species of plant and animal life on Earth. Given the variability among individuals, the long timescale provided by the new nineteenth-century understanding of geology was all that Darwin needed to explain the origin of all species from the original, primordial living cell. But – and it was a big but – what was the origin of that variability?

At the time the *Origin* was published, nobody knew anything about the origin of variation among individuals, nor the method by which characteristics were passed on from one generation to the next. The only working hypothesis Darwin had to go by was the concept of blending inheritance, that a new individual would inherit properties which in some way averaged out the properties of its parents. A large dog mated with a small bitch, for example, will typically sire medium-sized pups. But this blending process is the opposite of that required to maintain the variability of individuals. Within about ten generations, such blending ought to produce virtually identical individuals, with no variation for selection to act on. And why, if the process of heredity is simply a blending of the characteristics of both parents, are brothers so different? On the simple blending hypothesis, all full brothers ought to resemble each other as much as identical twins do.

This puzzle led Darwin and his contemporaries up at

least two blind alleys. It was clear that environmental factors during life could influence the adult form of an individual. Feed a baby well, and it grows up big and strong; feed it badly, and it will grow poorly and the adult may suffer from ailments such as rickets. In the absence of a better hypothesis, many scientists at the time were attracted to the idea that these environmental differences explained much of the differences between individuals, and furthermore it was widely thought that such acquired characteristics would be passed on to the offspring of the affected individual. The rickets sufferer, on this picture, would have children already predisposed to rickets, whatever their subsequent upbringing. Over the past century, however, many observations and experiments involving breeding plants and animals under controlled conditions have found no evidence for the inheritance of acquired characteristics in this way, and that hypothesis has therefore been rejected.

The other blind alley concerned the origin of the rest of the variation, that even in Darwin's day could not be explained by environmental differences. It was known that occasionally quite new features not present in one generation, or its known antecedents, appeared suddenly in an individual in the next generation. Such individuals were called sports – we would now call them mutations – and very often their new, previously unseen characteristics were passed on to subsequent generations. It's as if, to take a hypothetical example, white mice were bred in the laboratory for many generations, all of them white until one day a black mouse appeared in one litter, and all of the offspring of that mouse were black in their turn. This is a grossly oversimplified picture; but such sudden variations do occur and provided Darwin with another hypothesis, essentially that mutation provides enough variation among individuals in each generation for natural selection to operate effectively.

Very many people outside scientific circles still believe that this is what evolution is all about. Indeed, even within scientific circles there are experts who argue that such sudden changes from one generation to the next may be important, not in every generation but at some stages during the evolution of new species. More of this in the next chapter; first, however, it is important to appreciate that the kind of evolution described by Darwin does *not* depend on the occurrence of many mutations in every generation. Later investigations have established that mutations are far too rare to carry out this evolutionary role; the variability of individuals in every generation has quite another origin, one which Darwin never knew anything about, even though the pioneering work on this mechanism was carried out while he was still alive.

Variation and selection, plus inheritance, together account for evolution. Darwin explained selection, but he never provided a satisfactory explanation of variation, or inheritance. That came later, initially from the work of an obscure Moravian monk, Gregor Mendel, in the 1860s. It was only in the twentieth century, however, that the two aspects of evolution were fused into one complete theory, sometimes called Neodarwinism, or more often referred to as the 'Modern Synthesis'. Darwinism is just one leg of the Modern Synthesis; the other is Mendelian genetics.

# CHAPTER TWO

# MENDEL AND THE MODERN SYNTHESIS

Like begets like, but imperfectly. That is the key to evolution by natural selection. The offspring of a male dog and a bitch are always dogs and bitches, not rats or canaries, or oak trees. But none of their offspring is exactly the same as either parent. To understand how evolution works, it is necessary to go beyond Darwin's realisation that this variety is the basis of evolution, and find out how and why this variety is produced. Gregor Mendel took the first step on the road to understanding the mechanisms of evolution, the road from Darwin to DNA. But his work was in many ways ahead of its time. It wasn't simply because Mendel was an obscure Moravian monk that his results failed to take the scientific world by storm. The scientific world wasn't ready to learn about the 'particles' of heredity – what today we call genes – in the 1860s. Genes, the machinery of evolution, operate on the smallest level of the organisation of life, within the hearts of individual cells. But when Mendel made his discoveries, nobody knew very much about cells, how they work together to produce a living organism, and how a complex multicellular organism such as you or I grows from a single cell, a fertilised egg. Those developments came only in the twentieth century, when Mendel's laws of heredity were rediscovered and set in their proper context. Nevertheless, since Mendel's laws provide the essential key to understanding Darwin's ideas on natural selection, it is a happy coincidence that the correct place for a historical description of Mendel's work is, indeed, immediately after a description of Darwin's great work.

There is only one biography of this 'obscure Moravian monk', *Life of Mendel*, by Hugo Iltis, originally published in German in 1924, and only translated into English in 1932. Fortunately, this is a very good biography; the lack of any others is primarily because there is very little historical material available – even less today than there was in the 1920s – and it is hard to see how anyone could do a better job of organising and presenting the scraps that we have. Iltis was a native of Brünn (now Brno), the town associated with the Augustinian monastery where Mendel spent most of his life, and tells how he read Mendel's classic monograph in the museum library while still a schoolboy, and totally failed to grasp its significance. In the early 1900s, when Mendel's work and name became widely known, Iltis was becoming a scientist in his own right. He gained a PhD, became a teacher in Brünn, and eventually fulfilled his ambition of becoming Mendel's biographer.

# Mendel's early life

Although he only mentions them in passing, the difficulties Iltis encountered in finding material for the biography are interesting and important. As Iltis explains, 'being a priest, he [Mendel] had to be extremely cautious in the utterance of his philosophical views'. Mendel never kept a diary; his letters reveal little about the inner man; and, holding strictly to his vows, he shunned all relationships with women. Today, it seems odd that anyone with a scientific bent should become a monk – not just any monk, but one who rose to become abbot of his monastery. But there was nothing odd about this even to a scientist of Iltis' generation, and in the middle of the nineteenth century, in the heart of Europe, such a career seemed even less exceptional.

At that time and place, religion was still very much a part of everyday life. Moravia in the nineteenth century was not at the forefront of progress. The region has no political identity today, and had little even then, lying across the borders of what are now Poland, Czechoslovakia and the two Germanies; bits of the region have changed political ownership several times in the past couple of hundred years. To the son of an impoverished peasant family, born in 1822 in the tiny hamlet of Heinzendorf, religious belief was virtually unquestioned and the priesthood would provide the only opportunity to lead a life of study.

Young Mendel, baptised Johann, was fortunate to be taught elementary science in the village school. This was not included in the approved state curriculum, but taught in the local villages at the behest of the lady of the manor, Countess Waldburg. His enthusiasm was further roused by stories brought back by two older village boys who attended a higher school in the township of Leipnik, thirteen miles away. Encouraged by his mother and a teacher in the village, Mendel followed the same path in 1834. A series of glowing reports over the next few years encouraged his parents in their efforts to scrape up money for young Mendel's school fees. But when he left high school in 1840 there was very little prospect of Mendel continuing with a life of study. His father had suffered a serious accident in 1838, and been forced to sell the farm. Mendel joined the Philosophical Institute of Olmötz, but had scarcely any means of support, suffered ill health and had to withdraw from the examination in 1841. He was only able to return to try again in the following academic year because his younger sister, Theresia, renounced her share of the family estate, modest though it was, in his favour. With this slender resource and his income from private tuition, Mendel completed his course of philosophical studies.*

In 1843, however, there was only one possible way in

---

* Theresia's sisterly devotion did not go unrewarded. Mendel later provided help for her three sons. Two of these nephews, Dr Alois Schindler and Dr Ferdinand Schindler, in turn provided some of the background material for Iltis' biography.

which Mendel could proceed any further with his studies. He had to have security, a profession in which he would no longer have to worry about earning a living. The only choice open to him was the priesthood, and he duly joined the Augustinians at Brünn, which is now part of Czechoslovakia, on the recommendation of one of the professors at the Philosophical Institute. He was admitted as a novice on 9 October 1843, and took the name Gregor.

Brünn was the capital of Moravia, and the monastery still a cultural focus not just for the town but for the whole region. But there was no opportunity at first for the novice monk to play a part in any of this. Always reported as exceptionally diligent and of excellent behaviour, he devoted himself to theological studies for several years. With many older monks dying over those years, Gregor Mendel was ordained subdeacon on his 25th birthday, 22 July 1847, deacon on 4 August, and became an ordained priest two days later, even though he did not formally complete his theological studies until 30 June 1848. But he was not a success as a parish priest, and was delighted when, in September 1849, an opportunity arose for him to become a teacher at the high school in Znaim, a country town in southern Moravia.

Mendel seems to have been a good teacher, but he had never obtained the formal qualification required at the time. When he entered for the appropriate examination in 1850, he failed disastrously, the examiners pointing out that although he was 'devoid neither of industry nor talent' he lacked experience and had had no opportunity to acquire exhaustive knowledge of his subjects. Their opinion was that he might, given the opportunity for such exhaustive study, eventually become fit 'at least for work as a teacher at lower schools'. With this slightly less than enthusiastic recommendation before him, the Prelate at Brünn decided to send Mendel to study at the University of Vienna, which he attended from 1851 to 1853. By May 1854, he was teaching at the Brünn Modern School, a technical school founded a year earlier, where he remained a 'supply' teacher until 1868, never, ironically, succeeding in passing the required examination to become a full staff member, although he took the examination at least one more time, in 1856. Coincidentally, that year marked the beginning of Mendel's major research work, at the age of 34, which continued until 1871. By then, having been elected Prelate of the Augustinian community at Brünn in 1868, his other duties left no time for significant further research.

# Mendel's peas

Mendel had a passionate interest in science, something still regarded with suspicion by many of his peers and superiors in the church at the time, and had to tread carefully with his work. For a time, he had kept mice, cross-breeding them and observing the way different characteristics were passed on from one generation to another; it may be that this was no more than a hobby, or it may be, as Iltis suggests, that such experiments with animals were too far beyond the religious pale to be continued for long. Whatever the reason, he turned to botany for his crucial series of experiments – not so surprising for a farmer's son who had seen the value of plant breeding in practical terms. It's worth noticing, too, that these experiments began three years before the first publication of the *Origin*, and that although Mendel was an eager reader of the scientific publications of his day, collecting all of Darwin's works as they appeared, he was already well on the way to his own discoveries before Darwin's work was published. By the middle of the 1850s, he had a good scientific education, the security that had been lacking in his youth, and enough spare time to devote to his research in the monastery garden.

Opportunity alone, however, would not have been enough for the progress Mendel made in the seven years after 1856. He also had the capacity for hard work which attracted comment in all his reports as a student and novice priest, and a diligent and careful approach to his studies. But Mendel's great achievement was even more a forerunner of twentieth-century science than is sometimes appreciated – the important thing is not the discovery of isolated facts, but the logical way in which the facts are interrelated both with each other and with an overall theory to account for them. Mendel looked for *statistical* relations among the offspring of the plants used in his experiments, numerical ratios that put the subject of heredity on a secure mathematical foundation. He used many plants, in order to get reliable statistics, and took great care to keep the seeds from each type of hybrid separate so that he could follow the influences of heredity not just on one generation but through many generations. This attention to detail and genuinely 'scientific' approach runs through all of Mendel's work. But for now let's concentrate on the key work which – eventually – made his name, the study of hybridisation in pea plants.

Mendel chose the pea plant for study with great care. He had experimented with several other plants, and found the pea ideally suited to his needs. First, he had varieties which had been carefully cultivated for several years to ensure that each was a pure line – they 'bred true'. In addition, he needed a type of plant that could be fertilised conveniently by the experimenter. By removing the stamens from his pea plants before they ripened, Mendel ensured that the plants could not fertilise themselves, but were dependent on pollen he provided by brushing them with stamens from another plant. Finally, he realised the importance of studying single, clearly distinguished pairs of characteristics in his plants. Instead of trying to account for all the variety among the plants at once, he took things one step at a time, and studied the occurrence of pairs of very clear properties in his peas. Some pea plants have purple flowers, others white; in some the peas themselves are yellow, in others green; and in some the peas are wrinkled, in others smooth. In all, he studied seven such pairs of characteristics.

Mendel used pure strains, with traits that could be distinguished unambiguously. He looked at all possible combinations of those traits (crossing 'wrinkled' mother with 'smooth' father, 'smooth' mother with 'wrinkled' father, their offspring with each other and each type of parent, and so on), and he analysed the results carefully in statistical terms, combining results from many plants to ensure reliability. These are the features that made his results so outstanding. It took seven years for Mendel to carry the work on peas through, but by 1863 it was essentially complete, and the results can now be summarised in straightforward terms.

Take one example. Mendel took plants from lines which always produced green seeds and plants from lines which always produced yellow seeds, and carefully cross-fertilised them. If blending inheritance really did provide the basis of heredity in nature, the offspring ought to have been peas with an intermediate colour. In fact, the peas produced were all yellow, but when those yellow seeds were planted and the resulting plants allowed to fertilise themselves naturally, they did not breed true. In the next generation (the 'grandchildren' of the original two pure strains), three-quarters of the seeds were yellow and one quarter green (the actual numbers in Mendel's experiment were 8 023 peas, 6 022 yellow and 2 001 green. Each pod might contain five or six yellow and two or three green peas). So Mendel planted out these peas and waited to see what would happen. This time, the green peas bred true, producing pods containing only green peas, but the yellow peas produced a more complicated pattern. Out of 519 yellow peas, 166 produced plants in which all the peas in all the pods were yellow; the rest showed the same pattern as before, yellow and green peas in the ratio 1:3.*

---

* In fact, Mendel's published numbers are so close to the ratio 1:3 that they seem almost too good to be true. Some twentieth-century mathematicians have calculated that with only a few thousand peas in his 'sample', the chances of Mendel getting results that agree so precisely with the 'correct' ratio was only about one in ten thousand. In other words, if he had carried out exactly the same test 10 000 times, it would only have given the exact numbers he published once. Others interpret the statistics rather differently, and suggest that Mendel was guilty of nothing more devious than assigning doubtful peas (is this greenish-yellow one really yellow or really green?) to the category he knew they 'ought' to be in, on the basis of his years of work. Although it is impossible to prove the case, there is a suspicion here that Mendel may have tidied up his results.

However, since 1900 many, many more experiments of this kind have

From these studies, and much more work on the six characteristics he investigated, Mendel came up with a simple explanation. Each property (in this case, 'yellowness' or 'greenness') corresponds to a character which is carried by the pea. This character, which we would now call a gene, must be carried in a double dose in each pea, and one form of the character, or gene, dominates over the other, which is called recessive. If we call the gene for yellow **A** and the gene for green **a** (today, these two variations of the same gene are called alleles), then a pure-bred yellow pea can be represented symbolically as **AA**. A pure bred green pea is symbolised as **aa**. But, reasoned Mendel, when the two are crossed, the seeds produced inherit one version of the character – one allele – from each parent. The first generation of experimental peas, the ones which were all yellow, can therefore be symbolised as **Aa**. They are yellow because that allele, **A**, dominates the alternative, green allele, **a**. Only one allele in the pair is expressed in the form of the pea.

But what happens when these hybrids are allowed to produce offspring? Each new plant inherits one allele from each parent. Half the time, it inherits **A** from one parent and **a** from the other (**Aa**, on this picture, being the same as **aA**); one-quarter of the time, it inherits **a** from both parents, and is **aa** itself; one-quarter of the time it inherits **A** from each parent, and is **AA**. **AA** seeds, we know, are yellow. And **aa** seeds are green. But the **Aa** seeds are also yellow, because yellow is dominant. So altogether, in this second generation, three-quarters are yellow (one-quarter plus one-half) and only one-quarter are green. The same statistics explain the proportions of each type of pea in subsequent generations.

---

been carried out, and the pattern of behaviour reported by Mendel has been established to far greater accuracy. The interesting question of whether Mendel crossed the narrow line between polishing his data to show his evidence to best advantage, and actually cheating, can never be answered and is of academic interest only, because the established evidence for 'Mendelian' inheritance is today abundant and overwhelming.

All of Mendel's work pointed to the same conclusions, which are at the heart of the modern understanding of evolution. As well as studying individual characteristics, he compared more complicated hybrids, such as those produced by crossing plants with wrinkled-yellow seeds with those having smooth-green seeds, and showed that the different characteristics are inherited independently. His work established that organisms that reproduce sexually are 'constructed' in accordance with rules laid down by the genes. Each individual carries a double dose of genes, two alleles for each factor that is being described, though only one of these may be effective in determining the structure of the organism itself. This very important point distinguishes between the genotype, the 'blueprint' of instructions carried by the genetic material, and the phenotype, the overall physical appearance of an organism. The whole is not the same as the sum of the parts, because some of the parts are ignored. Peas which have a yellow phenotype may have one of two different genotypes, **AA** or **Aa**, whereas green peas always have the genotype **aa**, as far as those particular genes go. But an **Aa** genotype does *not* produce peas intermediate in colour, or striped green and yellow, as the idea of blending inheritance would suggest.

When sexual organisms reproduce, the gametes (pollen and seed in plants; sperm and egg in animals) each contain a single set of genes, with one member of each allele in the parent's genotype being segregated, at random, into each gamete. The *fertilised* egg, or seed, thereby obtains a full double complement of genes, one set inherited from each parent. But – and it is the most crucial 'but' in the whole story of evolution – the resulting genotype need not be identical to that of either parent, and will not be in practice, when we consider the vast number of characteristics that make up the overall phenotypes of most organisms.

Every organism – every phenotype – is now seen as the product of a large number of genes acting together, or sometimes in opposition, to produce the overall form. There are no intermediate forms at the level of genes – Mendel never found peas intermediate in colour between the yellow and green that he started with. But when a tall man and a short woman marry and have children, the children may grow up to be intermediate in size, because a great many genes together determine sex in the phenotype and their combined effect may come out somewhere in between the size of the two parents. Clearcut systems of dominant and recessive genes are not at all common, and in that sense Mendel's work involved special cases. This wasn't luck; although he didn't know about dominant and recessive characteristics when he started, it was the presence of those characteristics that made the pea plants what they were, and led Mendel to select peas to work with. The properties that made peas, as phenotypes, ideal for his experiments depended on the genotypes of the pea plants. It was only by understanding those special cases first, however, that biologists were able to advance to more general understanding of heredity and evolution.

# A prophet before his time

Mendel presented his results to the Brünn Society for the Study of Natural Science in February 1865. His audience was not greatly impressed, and Iltis, Mendel's biographer, speculates that the combination of mathematics and botany may have been both incomprehensible and slightly repugnant to the gentlemen of the society, all good scientists, but of the old school of natural philosophy. Mendel's careful planning, testing a specific model of reality with a series of logically constructed experiments, and analysing the results mathematically, seems natural to scientists today. In the 1860s, to most scientists it seemed incomprehensible. The very features that make us recognise Mendel's genius in the twentieth century made him a loner in the nineteenth.

Even so, Mendel's paper was duly published, along with the rest of the Proceedings of the Society for that year, in 1866. As a matter of routine, copies of the Proceedings volume were sent to well over a hundred other learned societies, who in turn regularly sent their Proceedings volumes to Brünn. Mendel's great paper, clearly setting out his discoveries, was thus available in academic libraries in London and Paris, Vienna and Berlin, Petersburg (as it then was), Rome and Uppsala. Nobody was struck by its importance; few people read it. But it did not escape the notice of the Church authorities in Moravia. Mendel came temporarily under a cloud in episcopal quarters, regarded (no doubt correctly) as a Darwinian. Pulling in his horns, he retreated into his monastic work, achieving great success and honour in the community, but never trying to promote his scientific ideas, and dying in 1884. When his paper was rediscovered, less that twenty years later, several of the scientists who eagerly went to the library shelves for the Proceedings of the Brünn Society for the Study of Natural Science for the year 1866 found that the pages of Mendel's contribution hadn't even been cut.

With hindsight, it is no surprise that Mendel's work failed to make an immediate impact. It *is* a surprise that he never wrote to Darwin about this work, and it is interesting to speculate as to how the understanding of evolution might have developed in the second half of the nineteenth century if Darwin had read Mendel's paper. But perhaps Mendel didn't dare to hazard his security by directly addressing his ideas to Darwin. And perhaps, even if Darwin had read the paper, the course of scientific history would not have been changed very much. The real problem, as far as getting these ideas accepted at that time was concerned, was that biology had yet to come up with any direct observations of the mechanisms by which Mendelian inheritance could work. Mendel's theory was abstract, based on mathematical reasoning. He gave names to invisible undetectable 'factors' which controlled heredity. The right time and place for such a theory was *after* microscopy had developed to study the inner workings of the cell, and to reveal the cell components called chromosomes.

To see how Mendel's ideas eventually did fit in to the mainstream of scientific progress, we will have to take one step backwards, historically, to find out how biologists began to understand the structure of the body, the nature of cells, and then the structure of cells. As we do so, however, it is worth remembering exactly what it was

Mendel discovered, five points which are, we shall see, directly related to the intimate structure of life.

1. Each physical character of an organism corresponds to one hereditary factor.
2. Factors come in pairs.
3. One, but only one, factor from each pair is passed on by each parent to each of its offspring.
4. There is an equal probability (in the strict, statistical sense) of either factor of a pair being transmitted in this way to any particular offspring.
5. Some factors are dominant and others recessive.

The story that fleshes out these abstract ideas, and gives physical reality to Mendel's factors, which we now call genes, is the story of the cell.

# Cells

The underlying structure of living things began to be appreciated in the seventeenth century, when the microscope was invented. In the 1660s, Robert Hooke published a description of his observations of plant tissues under the microscope, and he coined the term cells for the cavities, separated by walls, that he saw in thin slices of cork. But the cell theory in its modern form developed only in the nineteenth century, as ideas about the nature of life changed under the pressure of new discoveries and, especially, improving microscopic observations. It was only in 1838 that Matthias Schleiden, a German botanist, first proposed that all plant tissues are made up of cells, and it was a year later that Theodor Schwann extended the idea to animal tissues, proposing that *all* forms of life are based upon cells. During the 1840s, Schwann developed this theme. He pointed out that the cell represented the basic unit of life, that individual cells possess all of the attributes of life, and that all of the complex organs of living things, no matter how different their form and function may be on the large scale, are ultimately composed of cells. An egg of an animal, a seed of a plant, were seen for the first time as individual cells capable of reproducing themselves, dividing and growing to produce more cells which become organised into the adult form. Life could no longer be regarded as some mystic attribute of the whole organism, but was a property shared by the humblest cells.

'Each cell,' wrote Schleiden, 'leads a double life, one independent, pertaining to its development; the other intermediary, since it has become an integrated part of a plant.'* The same holds true for animals; as Schwann put it, the organism is a 'cellular state', in which 'each cell is a citizen'.

For a while, biologists were uncertain about the origin of cells. From his early observations, Schleiden gained the impression that cells grow like crystals, a kind of spontaneous generation, building up the cell structure from a central core. The full significance of the cell theory only really became apparent in 1858, when the studies of Rudolf Virchow showed that no cell could ever be seen coming into existence spontaneously. Wherever a cell exists, Virchow pointed out, there must have been a previous cell. In the same way that animals are only born of other animals, and plants arise only from the seeds of plants, so cells are produced only by the division of other cells. Life is never created on Earth today; all living cells are descended, in an unbroken line, from some remote ancestor in the distant geological past. Of course, the first cell, or cells, must have originated somewhere. But after 1858 there was no longer any mystery about the origin of the 'life' in each new plant or animal; each could be seen as the sum of those citizen cells, and each of those cells bore the stamp of life.

Virchow's realisation of the true nature of cells came in the same year that the joint paper by Darwin and Wallace

---

* Quotations are from *Schwann and Schleiden Researches*, trans. H. Smith, Sydenham Society, 1847. The development of the cell concept is discussed in more detail by Francois Jacob in *The Logic of Life*, and I have followed his treatment here. See also *Great Experiments in Biology*, ed. M. L. Gabriel and S. Fogel, Prentice-Hall, New York, 1955.

was read to the Linnean Society, a year before the publication of the *Origin*. The pieces of the puzzle began to fall into place, but it would take another generation of scientists, brought up on both the cell theory and evolutionary ideas, to probe within the cell and begin to understand the mechanisms of life and reproduction.

In Francois Jacob's words, 'with the cell, biology discovered its atom'.* The study of life became the study of cells. All cells are basically similar. They generally have sizes in the range from 10 to 100 micrometers, each one a bag of fluid, contained within a very thin membrane less than one-hundredth of a micrometer thick. The cells in which we are most interested are the ones that make up the structure of plants and animals like ourselves, and these all have a central dark nucleus. The cells themselves tend to be spherical in isolation, like soap bubbles, but, also like soap bubbles, they are squeezed and stretched into other shapes when joined to their neighbours. The cell wall, or membrane, holds each cell together as a separate entity, but allows chemicals to pass in or out of the cell, through the membrane, as required. The puzzle of life is the puzzle of how a fusion of one unusually large cell, the egg, with another, smaller cell, the sperm, can result in a single cell which proceeds through a series of complex stages to divide into first two, then four and eventually a great many cells, not at random but through a series of stages in which folds, buckles and indentations form and develop as the bundle of cells grows, eventually taking on the adult form.

In the second half of the nineteenth century, biologists knew, from the evidence of their own eyes, aided by the microscope, that the egg did not already contain a miniature human being, or hen, or cat ready to come forth and develop simply by growing larger. They could see the stages of development from the beginning for themselves, a controlled programme clearly following some master plan. What was the master plan, and where could it be concealed within the fertilised egg?

---

* *The Logic of Life*, page 121.

# Chromosomes

Every organism begins as a unit from the preceding generation, and for sexually reproducing species, such as ourselves, two members of the preceding generation are needed to make the basic unit. It is no mystery why like should beget like, because the offspring is developed from an actual piece of the parent. The variety which is so important to evolution must also be introduced at this stage, during the combination of hereditary information from both parents to make the fertilised egg (the same thing happens in plants, of course, but for convenience I shall concentrate on animals, with the human animal particularly in mind).

It is hardly surprising that after the middle of the nineteenth century the new science of cytology, the study of living cells, should have concentrated on how cells divide and reproduce. It took many people many years of research to fill in the details, but the key discoveries can be summarised very simply for the purposes of my present story. The crucial development came when Walther Flemming, a German anatomist, discovered in 1879 that the dyes used by cytologists to reveal structure within the cell are taken up very strongly by certain threadlike structures which became particularly clearly visible during the process of cell division. Flemming called the material which absorbed colour from the dyes so readily 'chromatin'; after 1888, the threads became generally known as chromosomes, and other bits and pieces of the cell became known as chromatids, chromoplasts, chromospires and so on. By killing cells at different stages during division, staining them with the dyes and examining the stained cells under the microscope, Flemming discovered the pattern and sequence of changes that go on in a cell during the normal process of division, a process which he called mitosis.

All living things grow because cells divide and multiply in this way. Many organisms, we now know, exist only as single-celled forms, and know no other way of life. Such a

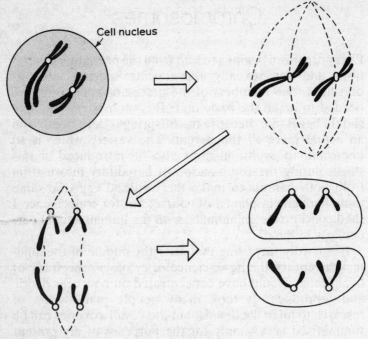

Cell nucleus

Figure 2.1 Mitotic division of a cell.

cell absorbs materials from its environment, processes them into the structure of the cell, and grows to a certain size, after which it divides into two and the two identical daughter cells repeat the cycle. Multicellular organisms use the same process of cell division to grow, and to repair damage or worn out tissues. It is going on in your body all the time, and it happens like this.

When a cell begins the phase of activity that will lead to its division, the first visible thing that happens is that the contents of the dark nucleus become arranged into the threadlike structures, the chromosomes. In fact, we now know that before the chromosomes become visible there has been a further phase of activity, inside the nucleus, during which each chromosome has been duplicated by the machinery of the cell. When they become visible, each chromosome looks under the microscope like a pair of identical threads, or chromatids, pressed together along their length but physically held together only at one point, called the centromere. In the next stage of mitosis, the chromosomes appear to get shorter and fatter, coiling up on themselves. Then, as the boundary between the nucleus and the rest of the cell disappears, a spindle-shaped structure, formed of tiny tubes, develops, stretching right across the cell from one 'pole' to the other. The ends of the spindle are structures called centrioles. The centromeres of the chromosomes become attached to the central part of the spindle – the 'equator', to continue the analogy – and are then pulled apart by the tubes. The centromeres divide in two and pull away from the equator, peeling the chromatids apart from one another so that one chromatid from each pair goes to each pole of the cell, each centriole. The chromatids are each now chromosomes in their own right. The spindle disappears, a new nuclear membrane forms around each set of chromosomes, the cell itself splits across the middle, the chromosomes themselves become diffuse and indistinguishable once again, merging into a new cell nucleus in each daughter, and the result is two new cells, each with a complete set of chromosomes identical to those of the original cell. Typically, the whole process takes some tens of minutes (see Figure 2.1).

During mitosis, the full complement of chromosomes in the original cell has been copied and passed on. That is how it is possible for each daughter cell to have a full set of chromosomes. Obviously, the chromosomes must be important to the cell, and it didn't take long for people to appreciate that this nuclear material must provide the guiding force, or blueprint, for the working of the cell. The whole point about mitosis is that it provides very accurate copies. Indeed, it is impossible to say which of the two cells produced by mitosis is the original and which the copy; both are daughters of the single parent that was there before. But this could not be the only form of cell division.

August Weismann, a zoologist based in Freiburg, in Germany, developed the idea that the germ cells of animals – the egg and sperm cells – must contain some essential requirements of life, that were passed on from one generation to another. He published the idea in a book in

1886, calling this mysterious 'something' germ plasm, to distinguish it from the ordinary body cells, known as somatoplasm. In the early 1890s, Weismann decided that the material of heredity must, in fact, be carried by chromosomes – the foundation stone of the modern understanding of heredity. His ideas were in many ways vague. There was very little evidence to go on in the 1890s. But Weismann did appreciate one crucial point. If the hereditary substances from both parents – the chromosomes – are mixed in the fertilised egg, it ought to contain twice as much hereditary material as the cells of the parents. In succeeding generations, the amount of hereditary material would double up impossibly. The only resolution to this puzzle, Weismann saw, would be if the germ cells were produced by a special process, a process of cell division which reduces the amount of hereditary material – the number of chromosomes – in half. This process of reduction division is called meiosis, and although we are getting a little ahead of the story in strict chronological terms, it makes sense to describe it now, alongside the description of mitosis.

In the special organs associated with the production of gametes – testes in males, ovaries in females – cells undergo a division process crucially different from mitosis. When the mother cell begins its activity, chromosomes are copied and then become visible, as they do in mitosis. For a particular organism, the same number of chromosomes is now visible as during the equivalent stage of mitosis, but each one appears on its own, not divided into two separate threads. Instead, the single chromosomes line up in pairs of very similar size (there is always an even number of chromosomes in the cells of living organisms, although the number varies considerably from one organism to another). Each chromosome now splits into two chromatids, joined together along a short part of their length, so that each pair of chromosomes becomes four tangled threads, and in a process very similar to the equivalent phase of mitosis a spindle forms and the threads separate. The crucial difference is that whereas during mitosis paired centromeres break and chromatids are separated, during meiosis the two threads from each of the original chromosomes stay together – the centromere itself does not divide – and the chromosomes themselves are separated from one another. But, as we shall see, something has happened to them before they separate.

After division, each of the two daughter cells formed undergoes a further phase of division, similar to mitosis but without first duplicating the chromosomes, so that altogether four new cells are produced, each with half the original complement of chromosomes, just as Weismann predicted. In males, three of these four cells usually become sperm; in females, only one of the four develops into an egg (see Figure 3.1, page 64).

The important difference between mitosis and meiosis is that mitosis makes exact copies of cells with a full complement of paired chromosomes – diploid cells. Meiosis produces cells which are not identical to one another, and which each contain only one chromosome of each type – haploid cells. When two haploid cells, egg and sperm, meet and fuse a normal diploid cell with a full double complement of chromosomes is produced. To anybody familiar with Mendel's work, the implications are obvious. Paired chromosomes are being segregated; one of each pair from each parent is combined in the new individual. It is only a small step to associate Mendel's factors with chromosomes, and, indeed, once Mendel's work was rediscovered it wasn't long before that step was taken.

# Mendel rediscovered

By the end of the nineteenth century, biologists were aware of the existence of chromosomes and suspected their role in heredity. Various theories about how the mechanism might work were being aired, and the obvious way to test them was by the kind of experiments Mendel had carried out more than thirty-five years before. So it is no surprise at all to discover that in the first years of the twentieth century several researchers were carrying out this kind of work, nor that some of them were using pea plants in their studies, for the same reasons which made the pea plant so useful to Mendel. The 'rediscovery' of Mendel came in March 1900, with the publication of two papers on hybridisation in plants by Hugo De Vries, of the Netherlands.

One of these papers, published in French, is very short, and contains no direct mention of Mendel, although the numerical results provided exactly agree with the Mendelian ratios of heredity. The other, longer paper appeared in a German journal, and goes into more detail of the theory behind this work. It also gives due acknowledgement to Mendel, and referring to his classic paper says 'this important monograph is so rarely quoted that I myself did not become acquainted with it until I had concluded most of my experiments, and had independently deduced the above propositions'.* If De Vries must have felt some frustration at finding his work anticipated by Mendel, imagine the feelings of Carl Correns, a German botanist, when he received a copy of De Vries' French paper. Correns, too, had been carrying out hybridisation experiments (some of them with peas), and he, too, had thought he had made an original discovery about the nature of heredity, only to discover, on diligently searching through the scientific literature, that Mendel had been there before him. And then, before he had published his own results, he was beaten to the post by De Vries as well. An Austrian, Erich Tschermak von Seysenegg, also discovered Mendel's work in much the same way, at about the same time and after achieving the same results independently. Then the floodgates burst open and confirmation came in from the United States, England and France. By the end of the year 1900, Mendel's place in scientific history was secure.†

---

* Quoted by Iltis, page 304.
† At least one historian of science has argued that Mendel was not really 'rediscovered', and that although his work did not gain widespread notice in the 1860s it was not completely ignored. Augustine Brannigan argues that Mendel's 'revival' in 1900 was largely due to a dispute between Correns and De Vries about scientific priority, with Mendel achieving his exalted position as a neutral, and dead, researcher more

The next significant step in understanding the process of heredity and evolution was taken by William Sutton, of Columbia University, in 1902. He wondered why the chromosomes that pair up during the early stages of meiosis should be physically similar to one another. Obviously, chromosomes come in pairs, and those pairs are separated during the later stages of meiosis. Only one member of each pair is passed on by each parent to its offspring. Sutton realised that the similar chromosomes parted in this way during meiosis must be replicas of the originally separate chromosomes that had come together in the fertilised egg, one set from each parent, when the new individual was created. In each chromosome pair, he reasoned, one member came originally from the mother and the other from the father, although each had been faithfully copied many times by mitosis since the egg from which the organism had grown had been fertilised. With Mendel's ideas current at the time, everything fell into place. The mysterious factors which Mendel had invoked to carry hereditary messages must be associated with chromosomes.

Each chromosome must carry many genes, since there are far too few chromosomes in the cell (23 pairs in man) to account for all of the Mendelian factors expressed in the phenotype. But here, at last, was the mechanism of Mendelian inheritance made visible. As long as alleles – different versions of the same gene – always occurred on homologous chromosomes, ones which paired and were separated during meiosis, Mendel's discoveries could be

acceptable to each of them than the other. ('The Reification of Mendel', *Social Studies of Science*, volume 9, page 423, 1979.)

Brannigan seems particularly puzzled that Mendel failed to promote his discoveries, and suggests he did not realise their importance. Altogether, it's the kind of argument sociologists love to stir up, but I don't think it holds water. It is no puzzle that Mendel kept a low profile, given his position in the Church and need for security; his classic paper was *not* widely reported prior to 1900: and, if it was 'really' well known to scientists, how come its pages remained uncut on the library shelves for more than thirty years? True, the paper was cited a few times prior to 1900; after 1900, however, Mendel became almost a household name, and Mendelism a familiar scientific term.

understood in simple, physical terms. This discovery was a great conceptual development. Before Sutton, the gene was an abstract idea, a mathematical entity required to explain the patterns of heredity. After his work, the gene had physical reality. It wasn't quite possible to see individual genes, but, with the aid of the microscope and suitable dyes, biologists could see collections of genes, chromosomes, at work. But how did these discoveries relate to Darwin's abstract ideas concerning the mechanism of evolution by natural selection? Ironically, at first the rediscovery of Mendelian genetics was seen as a body blow to the established Darwinian theory of evolution.

The Darwinian theory was based on the idea of small, gradual changes in an organism which accumulated as a result of selection pressure and produced all of the variety of evolution. But the Mendelian type of experiments all involved relatively dramatic changes in populations from one generation to the next. Of course, that is why the geneticists had chosen those particular organisms to work with. But in the early twentieth century a great deal of direct, experimental evidence of sudden changes accumulated, while the supposed process of gradual change postulated by Darwin remained undetectable – too slow to be seen. De Vries, in particular, challenged the idea that evolutionary change is a result of the accumulation of almost imperceptible differences from one generation to the next, and espoused the idea that evolution occurs as a series of giant steps, mutations which produce offspring differing dramatically, in some important characteristic, from their parents. Natural selection, the early geneticists believed, operated only to weed out the most obviously defective individuals after such a mutation, or evolutionary leap.

To anyone born after the Second World War, and acquainted in a general way with modern scientific ideas, it is salutary to read what Ledyard Stebbins, one of the great masters of modern evolutionary thinking, has to say about the place Darwin held in science in the 1920s. In his book *Darwin to DNA, Molecules to Humanity*, Stebbins tells how he first became acquainted with Darwin's theory in

1926.* Both of his professors at Harvard told their students that natural selection was a completely inadequate theory of evolution. The standard text which the students were set to read, a history of biology by Erik Nordenskiöld, stated boldly that 'to raise the theory of natural selection, as has often been done, to the rank of a natural law, comparable to the law of gravity established by Newton is, of course completely irrational. . . . Darwin's theory of the origin of species was long ago abandoned.'

This wasn't just some aberration among American universities – Nordenskiöld was Swedish, and the same views were held widely among scientists. William Bateson, in his Presidential Address to the British Association in 1914, had presented much the same view of evolution, a view which could certainly be regarded as the established view in the 1920s, even though there was opposition to it. It was only in the 1930s that further research led to a true marriage between Mendelian and Darwinian ideas, a synthesis of ideas that forms the basis of the modern understanding of evolution. That modern synthesis depended on two things – studies of plants in which the pattern of genes is more complex than the simple either/or choices in Mendel's peas (but still less complex than most choices in most living organisms, including ourselves), and, surely in the true spirit of Mendel's work, improved mathematical studies which showed just how small mutations could indeed do the work Darwin had required of them, in very large populations.

## The modern synthesis

Most characteristics of most organisms are not inherited simply as a choice between two alleles, like the yellow/green alleles in Mendel's peas. I have already mentioned that people are not just either tall or short, but come in all shapes and sizes, phenotypes built up in accordance with

---

* Page 46. The Nordenskiöld quotation is from the same source.

the interacting instructions from a whole array of genes. The way to extend Mendelism towards the more general case was to find plants that were sufficiently complicated to show the effect of several different alleles at work on one characteristic, and this was done by a Swedish geneticist, Herman Nilsson-Ehle. He found that by crossing a variety of wheat which has red kernels with another variety having white kernels he could obtain five different types, one red, one yellow-white, and three in different shades of pink. This is much more like the pattern when a tall man marries a short woman, or when a dark man marries a fair-skinned woman. The offspring are intermediate between the two extremes. But, crucially, Nilsson-Ehle found that the numbers of individuals with each of the five shades obtained by crossing the two original strains of wheat exactly obeyed the statistical laws Mendel had worked out, applied now to simultaneous transmission of two pairs of alleles located on two different chromosomes.

Edward East, of Harvard University, carried out similar experiments on tobacco plants with short and long flowers. What appeared at first sight to be blending inheritance could be precisely explained as Mendelian inheritance involving several genes. Continuing his work, East decided to look for the mutations that Darwinian theory required. He developed a pure strain of tobacco plants which all grew up identical to one another under controlled conditions. Then, he raised many generations of these plants, all under the same conditions in a constant environment. Even though the plants all started out with the same genotype, and they were all grown in the same environment, after several generations each plant grew up to be a little different from its neighbour. East concluded that the changes must be due to small changes in the genes themselves, spontaneous mutations, not on the grand scale required by De Vries but exactly in the way required by Darwin. Variety appeared spontaneously, but in a small way. Could these small variations really provide the raw material of evolution by natural selection?

This is where the mathematicians come in to the story. At about the same time, in the late 1920s and early 1930s, four geneticists, each with a thorough understanding of mathematics, independently realised that the studies of small families of plants and animals carried out by Mendel and his spiritual heirs could not give an accurate guide to the effects of genetic variation on large populations. Whereas even Nilsson-Ehle's wheat plants came in only five different colours, in a large population of interbreeding organisms – people provide as good an example as any – there are many more genes potentially available to be expressed in an individual phenotype. *Any* gene which exists in the cells of any human being can be expressed in the phenotype of some member of the next generation; any allele carried by any woman might be paired in one of her children with any allele of the same gene carried by any man. The enormous numbers of people on the face of the Earth give you some idea of just how much potential there really is, in principle, for variety among the genotype of the next generation. This great variety of alleles, far more than the pairs of possibilities among Mendel's chosen peas, is called the gene pool. Although each individual may have no more than a pair of alleles defining some characteristic, there may be many more versions of that particular gene residing in the cells which make up other human bodies.

What this means is that there is a great deal of variety among people, or among individual members of other large populations. If something happens to the environment which makes one allele out of that whole gene pool advantageous, then individuals carrying that allele will be more successful. Of course, this scarcely applies to *people* today, because we control our own environment. But imagine the situation even a few thousand years ago. Suppose something happened to the Sun's radiation which made it difficult for people with blue eyes to see (not a very likely possibility!). In the natural, pre-technological state, people with blue eyes would be at such a disadvantage that they would fail to survive, and would have fewer children than other people. Within a short time, the blue-eye allele would disappear from the gene pool. Equally, if, for some unimaginable reason, blue eyes became an advantage, then people with blue eyes would be successful, live long and have many children. The allele would rapidly spread through the gene pool. Selection operates, very effectively, on individuals. But the effects of evolution are seen in whole populations, because of the way alleles spread through the gene pool.

The mathematics which showed how effectively alleles could spread through populations were developed by R. A. Fisher and J. B. S. Haldane, in England, Sewall Wright, in the United States, and S. S. Chetverikov, in the Soviet Union. Some idea of the power of selection can be gained from Fisher's calculation, published in his classic book *The Genetical Theory of Natural Selection*, in 1930, that if a new allele, produced by mutation from an old one, gives those animals that possess the mutated gene just a one per cent advantage over those that do not, then the new allele will spread through the entire population within a hundred generations. As Andrew Huxley commented to the Darwin centenary conference in Cambridge in 1982, 'this power is far beyond what one ordinarily visualises when thinking of natural selection'.* An advantage which in individual terms is too slight even to be noticed by human observers studying a wild population of animals is enough to ensure the success of a mutated gene.

Evolution, according to the modern synthesis, *is* about mutations, but the mutations – spontaneous changes in the genes being passed by an individual to its offspring – need only be very small to do their work. Presumably, such changes occur because on rare occasions chromosomes are copied imperfectly during meiosis. More of this later. A very large gene pool is shared among many individuals, and natural selection very effectively decides which alleles stay in the gene pool and which ones disappear from it. Thanks to sexual reproduction, a minor mutation occurring in the egg or sperm of one individual can easily spread through the gene pool, if it is advantageous.

This is no more than the barest sketch of the modern

---

* *Evolution from Molecules to Men*, ed. D. S. Bendall, page 10.

synthesis, but this is not primarily a book about how evolution works at the level of populations, or even individual organisms, but about the search for the mechanism of evolution, deep within the cell.* The time has almost come to abandon the broad picture, and the excitement of the evolutionary debate, and to focus down upon the chromosomes themselves. How are they copied? How do genes control the workings of cells and bodies? What is the basic chemistry of life? How, indeed, do genes mutate to provide the variety necessary for evolution? Before I attempt to show how the answers to these ultimate questions were found, however, I should, perhaps, mention the latest round of the evolutionary debate, which has echoes of the old arguments about mutation and gradual evolution and which sometimes gives headline writers an excuse to suggest that Darwinian theory is once again under attack.

## Up to date

The current model of evolution involving sudden *large* mutations is called 'punctuated equilibrium', and has been championed by Stephen Jay Gould, of Harvard University. Gould is an extremely good writer, as well as an able biologist, and the combination has ensured a wide audience for these ideas. In a very simplified nutshell, Gould, and others, argue that species remain essentially stable for very long periods of time, millions, even hundreds of millions of years. During that time, Darwinian selection involving small mutations operates to keep an individual

---

* The solid basis of the modern synthesis, which I have barely discussed here, came from Fisher's book on the mathematical side, rapidly followed by T. Dobzhansky's *Genetics and the Origin of Species* (Columbia University Press, New York, 1937), which was the first readable account, and then by Julian Huxley's classic *Evolution: The Modern Synthesis* (Allen & Unwin, London, 1942). These are all worth reading. Huxley is particularly good, and the need for such a book in the early 1940s shows just how long it took to establish the fusion of Darwinian and Mendelian ideas.

species precisely tailored to its way of life, ideally fitted to its own ecological niche. Small changes in the environment are indeed tracked by evolution through natural selection, but by and large the Darwinian mechanism, it is suggested, acts to keep things stable. Occasionally, however, something different happens. Suddenly, a new variation on the theme appears, and sweeps through the population. After millions of generations of evolutionary stasis, a species alters, very rapidly, into a clearly different form.

The reason for the term punctuated equilibrium is obvious. But it is far less obvious what is meant by a 'sudden' change, in the context of evolution and geological history. As Gould himself acknowledges, a change that appears instantaneous from the fossil record in the rocks may have taken thousands or millions of generations. The question this new version of an old idea raises is whether two kinds of evolution are needed to account for both the stability of existing species and the creation of new ones, or whether good old Darwinian evolution can alone do the job. The traditionalists, led by spokesmen such as Francisco Ayala, of the University of California at Davis, have responded by pointing out just how quickly the accumulation of small, Darwinian changes can produce a significant change in the phenotype of an organism. In his contribution to the Darwin centenary conference, Ayala mentioned experiments with the fruit fly *Drosophila*, favoured for study by geneticists because of its rapid breeding cycle and interesting chromosomes. A large population of these flies, derived from a single ancestral pair, was divided into two. Half were put in a warm room, half in a cooler room, and each population was left to evolve on its own. After 12 years, the average size of the flies kept at 16°C had become 10 per cent greater than that of the flies kept at 27°C. Breeding at 10 generations per year, the populations diverged at a rate of 0.08 per cent per generation. How long would it take, at this rate, before such changes produced populations that could no longer interbreed, and would be classed as separate species?

Turning to human evolution, Ayala pointed out the dramatic change in brain size from *Homo erectus*, 500 000 years ago, to Neanderthal man, 75 000 years ago. In this short time, cranial capacity evolved from 900 cc to 1400 cc. It looks dramatic; but if we allow evolution to proceed at the same rate as in *Drosophila*, 0.08 per cent per generation, and if we assume a rather long interval of 25 years between generations, then the whole change could have occurred in only 13 500 years, or 540 generations. What looks, even from our relatively close evolutionary standpoint, like a dramatic burst of evolution is, in fact, no more dramatic than everyday changes going on all the time in some populations.

This debate continues. You can find an up-to-date account, from both sides of the fence, in the contributions by Gould and Ayala to the Darwin centenary conference volume already mentioned. For what it is worth, my own view is that Ayala provides the better description of reality. It seems likely that species do indeed remain evolutionarily static as long as conditions remain unchanging, but what matters is that they have the capacity to evolve 'rapidly', as fast as 0.08 per cent per generation, when under pressure from a changing environment. The punctuation marks of evolution are not sudden mutations, but environmental changes which result in new conditions and thereby select new variations on the theme from the gene pool. Whichever way the debate turns out, however, the important point to stress is that the idea of evolution by natural selection is *not* under attack; all that is being debated is whether occasional, large mutations play a significant role in providing some of the raw material for natural selection to act upon, or whether 'ordinary', small mutations are sufficient to provide the variety that is, truly, the spice of life.

That variety, however, does not come from mutations alone. Sex is an essential ingredient in mixing up the gene pool and offering new combinations of genes to the cutting edge of natural selection. As we begin to focus our attention down upon the double helix itself, the first thing we discover is that there is a lot more going on during meiosis than initially meets the eye.

# CHAPTER THREE

# SEX AND RECOMBINATION

By taking the story of Mendel's work through to its incorporation with Darwin's ideas into the Modern Synthesis, the cornerstone of evolutionary understanding today, we have really got a little ahead of the way the story actually unfolded. The Modern Synthesis began to appear only in the 1930s, and became firmly established only in the following decade. But from the early 1900s onwards, while Mendel's work and the work of later students of genetics were being incorporated into mainstream biology one of the main lines of progress came from studies of chromosomes, and especially the changes that happen to chromosomes when sex cells – gametes – are being made during meiosis. Indeed, an understanding of what sexual reproduction is all about was an essential prerequisite to a full understanding of both Mendelism and Darwinism. Without that understanding of sex, many eminent biologists remained highly sceptical about both Darwin's and Mendel's ideas – not the idea that evolution occurs, but the proposals as to *how* evolution occurs. The key to understanding the mechanism of evolution, the proof that both Darwin and Mendel were providing a good guide to the real world of evolutionary processes at work, came with the discovery that chromosomes themselves, unlike genes, are not permanent structures. Chromosomes can be broken apart, and the pieces recombined into new combinations, new packages of genes. This is what happens during meiosis, and it ensures that genes are constantly being shuffled around, from one generation to the next, throwing up new combinations to be tested against the cutting edge of natural selection.

The key figure during this stage in the development of ideas about evolution and the mechanisms of heredity was Thomas Hunt Morgan, of Columbia University. Morgan, born in 1866, came from a notable family. His great-grandfather, Francis Scott Key, wrote the US National Anthem, his father served for a time as US Consul in Messina, in Sicily, and one uncle was a Colonel in the Confederate army. In 1904, Morgan became Professor of Zoology at Columbia, and began the research which was his major contribution to science, and for which he was awarded the Nobel Prize in 1933. Morgan was among the many biologists who were uncomfortable with the Darwinian theory because of Darwin's failure to provide an explanation of how inherited characteristics are passed on from one generation to the next. He objected to the Mendelian theory, then gaining ground, because it rested upon the existence of then entirely hypothetical 'factors' which were passed on in the sex cells from parent to offspring, and although he acknowledged the possibility that chromosomes might have something to do with heredity, as late as 1910 he was arguing that specific hereditary traits could not be carried by individual chromosomes.

## The fruit fly factor

By then, Morgan had been involved in breeding experiments with the tiny fruit fly *Drosophila* for about two years. This insect turned out to be ideal for many genetic studies, and has been widely used in laboratories all over the world throughout the twentieth century. Its name, *Drosophila*, actually means 'lover of the dew', but it is not dew but fermenting yeast that attracts them to rotting fruit. The first reason they were chosen for study is that they are easy to keep and breed. Each fly is only an eighth of an inch long, and they produce a new generation in two weeks, each female laying hundreds of eggs at a time. A colony of *Drosophila* can be conveniently kept alive and well in almost any old glass container, and in Morgan's laboratory they used half-pint bottles. All these were good reasons for Morgan to choose *Drosophila* to study; it was pure luck, however, that the species turned out to have only four pairs of chromosomes, which made the investigation of the characteristics that came to researchers' attention a great deal easier than it might have been.

One pair of those chromosomes, as in all sexually reproducing species, is of particular importance, both to the species and the story of evolution. Biologists had noticed during the 1890s that although the chromosomes in the cells of all individual members of the same species look much the same under the microscope, one pair differs markedly between males and females. In this pair, one member, called 'X' because of its shape, is the same in the cells of both males and females. The other member of the pair is either another X or a different chromosome, shaped like an X with one arm lopped off, and dubbed 'Y'. In most species, individuals whose cells carry the XX combination are female; those who carry the XY combination are male. (In some species, including birds, the pattern is reversed, and in some species there is no Y chromosome, the possibilities being either XX or an X on its own. But these subtleties don't affect the story.)

People, as I have mentioned, each have 23 pairs of chromosomes in their ordinary body cells. Twenty-two of these are superficially the same in both men and women, and are matched pairs, or autosomes. The 23rd pair is XX in women and XY in men. From the outline of meiosis given in the previous chapter, it is clear that any egg cell must contain one X chromosome, inherited from the mother. But an individual sperm cell will contain either an X or a Y, because it can inherit either chromosome in the original pair present in its parent. So when X egg and X sperm meet and fuse, the result is a new female of the species – a baby girl. When X egg fuses with Y sperm, the result is a little boy. This realisation provided crucial evidence in support of Sutton's hypothesis that chromosomes

are copied and passed on to the next generation as a result
of meiosis, and showed that chromosomes did indeed play
a part in heredity – if nothing else, they determine sex,
which is a pretty crucial feature of any organism's
phenotype.

Morgan's original interest in breeding *Drosophila* was to
watch out for large scale mutations, macromutations in
which a new individual very different from its parents
suddenly appears. This kind of mutation is sometimes seen
in plants, but extremely rarely in animals. What he found
was much more subtle, but no less important. In nature,
most *Drosophila* have red eyes (geneticists, with a fine sense
of the dramatic, generally refer to the naturally occurring
form as 'wild-type', although the image of a wild
*Drosophila* is rather incongruous). In 1909, a variation
appeared in one of the populations housed in one of the
half-pint bottles in Morgan's lab – a single, male
*Drosophila* with white eyes. Following the same approach
as Mendel with his peas, Morgan mated the white eye with
one of its normal, red-eyed sisters. All of the offspring
were red eyes, showing that the 'factor' causing white eyes
must be recessive. So the experimenters looked at the next
generation, just as Mendel looked at the 'grandchildren' of
his original peas, and found something very strange. In this
second generation, there were 2 459 red-eyed females,
1 011 red-eyed males, 782 white-eyed males, but *no* white-
eyed females. Further studies always showed the same
pattern, and Morgan was led to the inevitable conclusion.
Whatever it is that causes some *Drosophila* to have white
eyes, the 'factor' responsible must be carried on the X
chromosome. This factor is recessive, and in the second
generation females it is always dominated by the normal,
red-eye factor in the other X chromosome. Additional
experiments also showed that white-eyed males had a
higher mortality in the egg than red eyes, explaining why
there were relatively few of them surviving to be counted as
adults in these experiments, and hinting that more than
just the characteristic for white eyes was being inherited in
the same sex-linked package.

In a series of further studies, Morgan found several other

properties of fruit flies that were linked with the sex of the flies and must be carried on the X chromosome. He adopted the name gene, introduced in 1909 by a Danish botanist, Wilhelm Johannsen, for an individual Mendelian factor, and he concluded that Sutton was right, that one chromosome carried a collection of genes strung out like beads on a wire. The evidence of his own experiments with fruit flies filled in the gaps which were so glaringly obvious to Morgan in both Darwinian and Mendelian theory, and he convinced himself that both were, indeed, correct. This powerful example of the scientific method at work is all too often glossed over in the story of the development of evolutionary thought; a sceptic who convinces himself that the theory is correct as a result of his own experiments carries much more weight, it seems to me, than a disciple who eagerly swallows up the ideas of a great predecessor and regurgitates them unthinkingly. Scepticism is the foundation stone of science, and it was Morgan's work, more than any other single contribution, which established the basis of heredity and began the fusion of Darwinism and Mendelism by about 1910. But that was only the beginning of the story.

# Broken chromosomes

Morgan's work at Columbia developed in the second decade of the twentieth century in collaboration with his students A. H. Sturtevant, C. B. Bridges and H. J. Muller. The simple Mendelian laws of genetics apply only if genes are transmitted independently, and the discovery that genes are linked to one another by being physically associated on one chromosome or another resolved much of the confusion about inheritance that did not follow the simple laws worked out by Mendel from his studies of peas. As more minor mutations were discovered among the populations of *Drosophila* housed in the Columbia lab, several more were found which always went together, like maleness and white eyes. Each group of genes that is inherited together is called a linkage group, and Morgan's team found that there were just four linkage groups required to explain the pattern of inheritance in *Drosophila*, exactly matching the number of chromosome pairs. Later studies have always shown the same pattern in other species – never more linkage groups than there are pairs of chromosomes to carry them. But as enough information was gathered to make it possible to study the behaviour of linkage groups themselves, new anomalies emerged from the data.

One example should make clear the kind of new discovery that emerged. Wild-type *Drosophila* have grey bodies with long wings. One mutant type of fruit fly has short wings and black bodies. Both of these characteristics are recessive, and when the two types are interbred all of the first generation offspring have grey bodies with long wings. As usual, the interesting things emerge only in the second generation.

There seemed to be only two possibilities for the grandchildren of the original flies chosen for this experiment. If the two genes, for black body and short wings, are housed on separate chromosomes then they ought to be inherited in a straightforward Mendelian fashion, appearing in the second generation in the same predictable ratios as the occurrence of yellow and wrinkled peas in the monk's equivalent experiments involving two inherited factors. Perhaps, though, the two genes form a linkage group, and are carried by the same chromosome. In that case, the pattern in the second generation ought to be very simple, just as if one gene were being transmitted, with three-quarters of the flies grey with long wings. Just one-quarter, in that case, would carry the recessive version of both genes in both alleles, and have black bodies and short wings. That outcome would mimic the original Mendelian experiments with just yellow and green peas.

In fact, out of a large number of experiments of this kind, it turned out that the results are very close to the pattern expected if the two genes are part of a linkage group, but they are not exactly in line with this simple prediction of Mendelian genetics. A few flies appeared that had grey bodies and short wings; a few had black bodies and long wings. From many studies of this kind, Morgan was driven to the conclusion that linkage groups are not unbreakable entities. Sometimes, somehow, one or more genes on a chromosome change places with the equivalent genes – the alleles – carried by the chromosome's paired partner. The association is literally broken and, in this case, the two genes, for black body and short wings, are separated on to different chromosomes and transmitted independently to the fly's offspring. The only time this breakage and segregation can happen is during meiosis. Chunks of chromosome are broken apart, swapped between the pairs, and recombined into new arrangements of alleles. Each chromosome always carries a set of genes for a particular set of characteristics, but the alleles are reshuffled and recombined.

# Recombination

Since Morgan's day, uncounted numbers of breeding experiments of this kind have been carried out, and ever-improving microscopic techniques have provided direct observations of what happens when chromosomes break and recombine during meiosis. The trick happens during the earliest phase of meiosis, when the pairs of homologous chromosomes in the cell (one member of each pair inherited from the organism's father, the other from its mother) meet and tangle with one another. During this process, each of the original chromosomes divides into two chromatids, still joined by a centromere, which are duplicates of the hereditary material from the parent. So there are four threads which tangle and overlap before being pulled apart in pairs, still joined by their centromeres, to opposite ends of the cell. During this tangling, where the chromatids cross with their equivalent threads from the other chromosomes they can be physically broken apart, with the broken ends joining up with the broken ends of their opposite numbers. This need not happen just in one place but can occur at several places along the chromatids, with the result that when the threads separate each one is a mixture of bits from each of the original chromosomes.

Bits of material have literally crossed over from one chromosome to the other, and the process is sometimes referred to as 'crossing over'. But it is always reciprocal, with each piece of swapped material being replaced by the exactly equivalent stretch of the other chromosome. The more accurate descriptive name for the process in recombination. A useful analogy is to think of the two threads from one chromosome, the paired chromatids, as green string, and the two in the other pair as red string. After the tangling and separation stages of meiosis, there will be four threads which are each partly red and partly green. Everywhere a green section has been inserted into a red thread, the 'missing' piece of red thread has been inserted into the gap left in its opposite number by the removal of the green section. The result is that although in almost every cell of your body it would be possible, in principle, to pick out which chromosomes came from your mother and which member of each pair came from your father, in the sex cells there are new chromosomes which each contain a mixture of genetic material from both of your parents. It is these new chromosomes which you pass on to your own children, and which account for the astonishing amount of variability among individual members of species that reproduce sexually.

Crossing over appears to occur at random – chromosomes can, in principle, be broken and rejoined anywhere along their length, where the chromatids just happen to cross when they get tangled up during meiosis. If that is the case, Sturtevant realised, then genes which are further apart from one another on a chromosome are more likely to be separated during recombination, because there is more likelihood that the chromosome will happen to be broken in between them. In a series of brilliant experiments which he began in 1913, Morgan's pupil developed this simple realisation into a tool for mapping the chromosomes of *Drosophila*. Genes that are further apart can be separated more easily, and that means more often, in terms

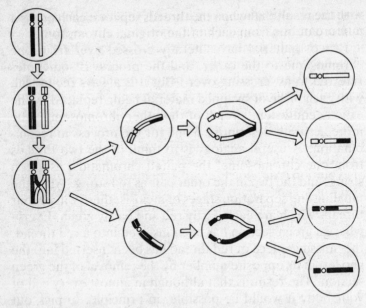

Figure 3.1 During meiosis, described on page 44 of the text, chromosomes are broken and rejoined to form new combinations of genetic material. This is called crossing over.

of generations; genes that are close together only get separated occasionally, when the break in the chromosome occurs just in between them. From a series of many breeding experiments and careful statistical analysis, checking just how frequently particular linkage groups of genes got broken, Sturtevant and his successors have used this behaviour to determine how far apart the genes on a particular chromosome must be, and the order in which the genetic 'beads' are strung along the chromosome's 'wire'. Today, seven decades later, more than 500 genes and their exact relationships to one another on the four chromosomes of *Drosophila* have been identified, and the chromosomes of many other species have been mapped in lesser detail. In 1915, Morgan, Sturtevant, Bridges and Muller published their results in a book, *The Mechanism of Mendelian Heredity*, which influenced a generation of geneticists, and remains in print to this day. That publication date is as convenient a marker as any to choose as the necessarily somewhat arbitrary date by which it can reasonably be said that the necessary combination of Darwin's and Mendel's ideas had begun to become clear. This was only a beginning; another generation of biologists still had to obtain the final proof of these ideas. But the evidence gathered by Morgan and his colleagues is, with hindsight, already clearly indicating that the process of variation which Darwin needed to provide the raw material of evolution is *not* due to significant mutations occurring in every new generation, but to a constant reshuffling of the genetic pack, which deals new combinations of alleles out to every individual (except for identical twins, produced when a fertilised egg divides into two), and which gives an opportunity for occasional, rare mutations to be tested along with the rest.

# Variety is the spice of life

The great thing about sex and recombination is that together they provide a means for different versions of a particular gene to be shared, in effect, by all of the members of a population. Although I have at most two alleles for any particular human gene, and in very many of my chromosomes each member of the pair carries copies of the same allele for many genes, my children could inherit any allele for any human gene that is carried by any fertile woman. Furthermore, my grandchildren will then inherit completely new chromosomes which will have been produced by a process of crossing over and recombination involving the chromosomes which my children inherit from me and the set they inherit from my wife; each grandchild's other set of chromosomes will have been formed from a comparably large pool of potential possibilities involving the genetic material of the parents of the person that my own child – the grandchild's parent – marries. Setting it out in words, it is clear that the number of possible potential combinations of genes in the chromosomes of any one human being must be enormous. Putting the numbers in produces quite incomprehensible figures, far beyond the range of even astronomical numbers.

Francisco Ayala has done the calculation for us, in his contribution to a special issue of *Scientific American* on evolution.* In human beings, only 6.7 per cent of genes are heterozygous, on average. That means that just over 93 per cent of the alleles carried by the chromosomes in your body cells are the same on each chromosome of a pair. The potential for variation doesn't sound too impressive yet, but wait. In round terms, there are about 100 000 genes which together go to describe a human being, and determine your phenotype – you are a much more complex animal than a fruit fly. 'Only' 6 700 of these genes will consist of different alleles on the paired chromosomes, so there are 'only' 2 to the power 6 700 ($2^{6700}$) different ways in which the chromosomes in your body or mine might recombine to form different, new chromosomes to pass on to the next generation. What does a number like that mean? Converting into the more familiar base 10 arithmetic, we get the no less incomprehensible number $10^{2017}$ possible new combinations of genetic material which I could pass on to my children, and you to yours. Let's call it $10^{2\,000}$. In his book *Cosmology*,† Edward Harrison, of the University of Massachusetts, makes a simple calculation of roughly how many nucleons (protons and neutrons – the particles that make up an atomic nucleus) there might be in all the stars and planets of all the galaxies in the Universe. The number was first calculated by the great astronomer Arthur Eddington, in the 1920s, and present-day calculations come out with much the same sort of figure, based on various reasonable assumptions. In round terms, it comes out as $10^{80}$.

Increasing powers of ten very quickly run away with themselves, and the difference between $10^{80}$ and $10^{2\,000}$ is

---

* September 1978, volume 239, number 3. Ayala's article 'The Mechanisms of Evolution' is on pages 48-61.
† Cambridge University Press, 1981, page 343.

vastly greater than the difference between 80 and 2 000. Even $10^{82}$ is 100 times bigger than $10^{80}$. To get $10^{2\,000}$ you would multiply $10^{80}$ by $10^{1\,920}$, or to put it another way to write $10^{80}$ as a percentage of $10^{2\,000}$ you would have to set down a decimal point followed by 1 918 zeroes and a 1. The total number of all the protons and neutrons in the Universe is an utterly insignificant number compared with the number of different possible combinations of genetic material that could be produced by *one individual* human being. Perhaps that gives you a feel for the potential variety of human genotypes, and therefore the variety of phenotypes, that might be produced to be tested by natural selection – which operates, of course, on the individual members of a species. Today, civilisation insulates us from natural selection to a large extent, but the same numbers applied to our ancestors, and also apply, in round terms, to all other large mammals.

And although only 6.7 per cent of genes (or rather, gene loci) are heterozygous in each human being, there may be many more than just two alleles of a particular gene present somewhere in the human population. The whole gene pool is available to create new chromosomes, new genotypes and new phenotypes. Some combinations turn out to be very good, and the resulting individuals thrive and, crucially as far as evolution is concerned, produce many offspring in their turn. Some are relatively poor phenotypes, and produce fewer offspring. In effect, natural selection is picking out the best combinations of alleles, the fittest linkage groups. But such is the potential for variety that even alleles which are not the best possible versions of a particular gene for survival in a particular environment – at a particular time and place – still exist in a few individuals. When conditions change and those alleles suddenly become much more valuable, the individuals that carry them will do well and thrive, producing more offspring than those individuals which carry a different allele, or linkage group, affecting that particular characteristic. Many years later, looking back on the fossil record, paleontologists may see a geologically dramatic change in

the species, and start talking about punctuated equilibrium. All that need have happened, however, is that the
environment changed and selected preferentially a different combination from the vast genetic pool of the
species. No longer is it any surprise, either, that Darwin's
finches should have become so superbly adapted, in a short
time by geological standards, to the particular ecological
niches they occupy on the different Galapagos islands. All
of this can happen without any mutations occurring at all.
If and when mutations do occur, they may have little
immediate impact, but produce yet another variety of a
particular gene, an allele which confers no special, immediate advantage on the individuals that carry it but which lies
around in the gene pool and only becomes prominent when
conditions are right for it. *

One interesting implication of all this is that neither
mutations nor recombinations must happen too frequently, or potentially beneficial new alleles and linkage
groups would be destroyed before they had a chance to
spread through the gene pool and to be tested in different
combinations with other genes in different recombinations
of chromosomes. Many studies show that in all living
organisms on Earth, from man to maize, fruit fly to
bacteria, the rate of mutation is roughly constant, with a
change occurring spontaneously once in every gene pair in
every generation for every 500 000 individuals in the population. R. A. Fisher showed mathematically that this slow
rate of mutation, combined with Mendelian inheritance
and random recombination of chromosomes into new linkage groups, is exactly what is required to explain evolution
by natural selection. But that, of course, is putting the cart
before the horse; what has really happened is that the
process of mutation and recombination has itself evolved,
over millions of generations, to fit the most effective pattern. Individuals that 'used' recombination in any way that

---

* The relationship between sex and evolution, with special reference to
our own species, is discussed more fully in *The Redundant Male*, which I
wrote with Jeremy Cherfas.

was less efficient than the way it is used today lost out in the struggle for survival, and have left no heirs; only the fittest survive, and the mechanism of heredity is by no means excluded from the selection process. But although the role of mutation is much less than Darwin thought, it is, nevertheless, a crucial role. Studies of crossing over and recombination help to explain how some mutations occur, and set others in their proper context.

# Mechanisms of mutation

Mutations – errors in copying the genetic material – are the ultimate source of the variation which provides the basis for evolution by natural selection. If there had been no copying errors, all living things on Earth would be exact replicas of the first living thing that appeared on Earth. The copying process must be accurate enough to ensure that like begets like. Indeed, errors must be rare. But they must occur just often enough to provide the raw material of evolution, and the rate at which mutations occur has itself been selected by evolution over thousands of millions of years. The nature of these copying errors did not become clear for many decades after the work by Morgan and his colleagues, from investigations of DNA, the double helix itself. But we can still understand in general terms what goes on when mutations occur, using language which would have been intelligible to an up-to-date geneticist of the 1920s.

Chromosomes carry information which is used to construct and operate the phenotype of the organism whose cells the chromosomes inhabit. Without knowing exactly how that information is communicated and used by different parts of the organism, we can see that the whole package of genetic material in the chromosome acts as a blueprint for survival, describing what each part of the organism should do, and how it should react to different circumstances – different stimuli from the environment. The image of genes as beads strung out on a wire already carries an important clue that this coded message of life is written in a linear fashion, like the lines of letters and words which make up the message you are reading in this book. Just how precisely apposite this analogy is we shall see later; for now, let's take it at face value and look at how a message might get changed when it is being copied.

The simplest kind of 'mutation' would be if a letter got left out, or an extra letter was inserted. Think of a typist copying out the sentence 'The cat sat on the mat'. All of us who have learned to type remember how easy it was, before we became skilful, to produce variations such as 'The at sat on the mat', or 'The ccat sat on the mat', or 'The at sat on the mmat'. These are point mutations, changes occurring at one point in the message, deletions or insertions of a single letter. A third kind of point mutation, substitution, would change one letter into another, with 'The cat sat on the mat' becoming, perhaps, 'the cat gat on the mat'.

Any random change in the genetic message is almost certain to make the message, or part of it, into nonsense. Such a mutation is clearly harmful, and will result in a phenotype that does not work as efficiently as its competitors, and which will lose out in the Darwinian struggle for survival. But sometimes such changes produce new sentences that also make sense, even if not the same sense as was in the original message. Change 'The cat sat on the mat' to 'The cat sat on the hat', and still the message is telling us something. Very occasionally, this kind of change in the genetic code might produce a phenotype which did some small thing more efficiently than its rivals. And a small advantage is all that evolution needs to work with.

Clearly, though, it would take an enormous amount of time for these simple point mutations to build up new varieties from an original single-celled organism with a simple genetic blueprint. Indeed, evolution did proceed extremely slowly on Earth for thousands of millions of years after single-celled organisms appeared. The whole process speeded up enormously once sex and recombination were 'invented' – at least, it did for those species that use sex; the almost unchanged descendants of those original single-celled organisms still exist, unchanged, in vast profusion on our planet.

Recombination introduces the possibility of new kinds of mutation. I said before that during crossing over each piece of chromosome is swapped with a precisely equivalent piece of the paired chromosome, but that isn't always quite true. It shouldn't be any great surprise, if you think about the nature of this copying process, to learn that sometimes a piece of chromosome gets turned end for end before it is inserted into the proper place. In our little message, 'The cat sat on the mat' becomes 'The tac sat on the mat', which doesn't make much sense at all. Again, most such mutations are harmful. But try a variation on our message, making the same kind of change to the sentence 'The dog sat on the mat'. Now, the copying error – inversion – produces the message 'The god sat on the mat', which does make sense. Inversions, like point mutations, can very occasionally be good news in terms of the working of the phenotype.

Even the ordinary copying process of mitosis has more scope for producing mutations than simple deletions or substitutions of single 'letters'. The mechanisms which copy chromosomes can occasionally make the kind of errors a tired typist or typesetter makes, copying the same 'words', even whole sentences, twice, or leaving a sentence out altogether. In a multicellular species like ourselves, such a change in the odd body cell is of no significance whatsoever. But if it happens during the copying stage of meiosis, following recombination, then there is a chance that the modified chromosome will be passed on to the next generation. Starting out from a single sperm or egg, it will be incorporated into every cell of the new individual, and be passed on to future generations, if the individual survives and reproduces. An extra copy of a gene, or part of a gene, may serve no purpose at all, but will still be copied faithfully in future cell divisions, passed on from generation to generation and carried like excess baggage in the cell. Although the full realisation of the potential this offers for evolution only became apparent in the 1970s and 1980s, such 'spare' bits of genetic material can themselves mutate, passing through several, or many, useless variations until they happen, by chance, to fall into a pattern that does something new and beneficial for the cells they inhabit. A whole new gene might be created in this way. But that really is a story which will have to come later, in its proper historical context.

So the copying process can introduce new bits of material to the chromosomes, it can scramble the code that is already there, and it can, of course, leave a chunk out if the broken ends join up across a gap during meiosis, leaving a piece of chromosome out on its own. The fragment may get lost, failing to be copied as meiosis proceeds, or perhaps this is how new chromosomes are produced. Sometimes during meiosis pieces of material are exchanged between chromosomes that are not homologous pairs, and again new chromosomes may appear as a result. But there is nothing to be gained in cataloguing all the possibilities. The rule of thumb is that if you can imagine a way in which a linear message might get corrupted during copying, then it probably happens, if only occasionally, during both meiosis and mitosis. And a personal example may help to emphasise how important these simple copying errors can be. Human beings, as I have mentioned, carry 23 pairs of chromosomes, including the sex chromosomes. The gorilla and chimpanzee, our closest relatives, each have 24 pairs of chromosomes, and the similarity between the chromosomes of the three species is so great that the equivalent pairs in each set can be identified. One pair of human chromosomes has been formed by the fusion of chromosomes very similar to those in two pairs of chimp or gorilla chromosomes. But modern techniques can reveal more subtle details than this. The main further difference between man and chimp is that the set of human chromosomes carries six inversions, compared with those of the chimp; we are separated from the gorilla by eight inversions. It is actually possible to see under the microscope the differences between the chromosomes of ourselves and our nearest relatives, the differences which, as a result of a very few mutations, have resulted in the presence of three species on Earth, descended from a common ancestor which was around (according to the best evidence) about five million years ago. Something to ponder next time you watch the chimpanzees' tea party at the zoo.

## Doubled chromosomes

There is still another form of mutation, however, which, although it doesn't affect human evolution directly, is of the greatest importance to the survival of humankind. If a cell fails to divide after the chromosomes have been copied, the result will be a cell containing a double set of chromosomes. Subsequently, the cell may go through a normal divison, copying the doubled set of chromosomes and then dividing to produce more cells with the extra genetic material. A cell with the normal set of chromosomes, in pairs, is called diploid; a doubled set produces what is called a tetraploid cell, or organism. In sexual reproduction, just as ordinary diploid organisms halve their chromosome complement to produce haploid gametes, so a tetraploid organism halves its chromosome complement to produce diploid gametes. A diploid gamete can combine with a haploid gamete from the organism's sexual partner, and the resulting hybrids (which are not at all common) are triploid organisms carrying three sets of chromosomes. But for practical purposes they can be ignored, because there is no way for the chromosomes to pair off during meiosis. Such hybrids are, therefore, sterile and produce no offspring of their own. That means, for example, that if a mutation produced a tetraploid human being then even if he or she became the parent of a child (which is not at all likely, but possible) that individual would leave no offspring, because the child would be infertile – the triploid cells simply could not produce functioning gametes through the usual process of meiosis and recombination.

In plants, though, things are rather different. Plants are often parthenogenetic. One individual organism produces both male and female gametes, and these can fertilise each other to produce new plants, with many of the benefits of sex and recombination but none of the hassle of finding a partner. A single abnormal cell division might occur at a crucial stage in the development of a plant, perhaps at the point where a new bud is forming. That bud may develop, as a result, from repeated division of the descendants of a single tetraploid cell, into a whole tetraploid branch carrying tetraploid flowers and tetraploid pollen. Seeds produced by self-fertilisation of those flowers will develop as complete tetraploid plants. In effect, this creates a new species, because the tetraploid plants will only be able to reproduce among themselves. This can happen repeatedly, and hybridisation between different species can also change the chromosome number, producing varieties with many different multiples of the original chromosome set. The phenomenon is called polyploidy. But the plants will still be recognisably members of the same family, with chromosome numbers that are multiples of some basic set. Even a hybrid that might seem to be infertile, because it carries a triploid set of chromosomes, for example, can itself produce a new fertile strain if it too undergoes the doubling process, producing, in this case, a variety in which the gametes carry three sets of chromosomes and the adult form has six.

The picture is just as complicated as it sounds. It is quite common in nature, and it can be triggered in the laboratory by treating plants with an extract from the autumn crocus. Polyploid plants are generally bigger overall than their 'normal' counterparts, just as you would expect for individuals in which every cell is bigger than it 'ought' to be. They have thicker leaves, larger flowers and fleshier fruit, and bigger seeds. That is why they are, and have been, of crucial importance to humankind. Polyploid plants provide better food. The earliest wheat cultivated by man was a diploid species, with 14 chromosomes (seven pairs). Modern wheat, used in making bread, is a hexaploid form with 42 chromosomes. That is one of the main reasons why modern wheat plants are so big and produce such large seeds.

Polyploidy must have played an important part in increasing the amount of genetic material from the relatively tiny quantity found in the cells of bacteria to the many pairs of large chromosomes in our own cells. Mutation as a whole aids natural selection in many ways, both providing the raw material of evolution and ensuring that new species split off from old stock. A relatively minor mutation which does not have a profound effect on the phenotype may produce a variety which can no longer reproduce with the original stock, as in the case of most polyploid varieties, and from then on evolution is selecting from not one species, but two. The whole thing is so beautifully balanced that it sounds almost too good to be true. Can nothing go wrong? Unfortunately, it can. Some genes, not to put too fine a point on it, cheat. But the way in which they cheat also reveals a lot about how evolution works.

# Genes that cheat

Genetic cheating takes advantage of recombination to give particular genes an edge by ensuring that their rival alleles are not passed on to functioning gametes, and so do not become part of the genotype of the offspring. It first came to light in experiments at the University of Wisconsin in the mid-1950s, where Yuichiro Hiraizumi was studying *Drosophila*. These studies, like those in Morgan's lab that provided some of the key early insights into recombination, involved genes affecting eye colour in the flies. The 'wild-type' flies have eyes in which there are two pigments, one bright red and the other brown, which together give them a characteristic dark red colour. One mutation produces flies in which there is only red pigment in the eyes; another mutation eliminates the red pigment and leaves brown; and when both mutations are present in the same individual the eyes are both white, without either pigment. Both of these characteristics are recessive, and have no effect if the corresponding normal allele is present on the appropriate paired chromosome. And both are carried on the same chromosome.

Hiraizumi's tests involved mating wild-type *Drosophila*

with different pure-bred laboratory strains whose chromosome patterns were well known. In the first step, red-eyed
males were produced, with (ignoring all the chromosomes
having no effect on eye colour) one 'wild' chromosome and
one from a laboratory strain of fly. Now, it happens that
crossing over does not usually happen during meiosis in
male *Drosophila*, which is yet another advantage for geneticists studying mutations – it gives them one less thing to
worry about. So when these hybrid males were mated to a
laboratory strain of white-eyed females, carrying two
mutant chromosomes, Hiraizumi expected that the offspring would only inherit either the wild chromosome with
both 'good' eye-pigment genes or the laboratory chromosome, carrying the red-eye package, from their fathers.
There would be no mixing of the paternal genetic material.
In that case, the offspring of this second mating ought to be
half normal, with dark red eyes, and half white eyes.

Out of roughly 200 matings, that is exactly what
Hiraizumi found – *almost* all of the time. From six of the
matings, however, he found something totally unexpected
– more than 95 per cent of the offspring had red eyes.
Something very odd was happening, and what's more the
*same* odd thing had happened six times to six separate (but
equivalent) chromosomes inherited from six separate
males in the original wild population. Later studies showed
the same phenomenon among a small proportion of
*Drosophila* in every wild population. Through a series of
further experiments, Hiraizumi and his colleagues established that what was happening was that the chromosome
carrying the red-eye package of genes was doing something
during the pairing-up phase of meiosis that altered its
paired chromosome and ensured that sperm carrying the
normal chromosome could not function properly. Males
carrying this mutant chromosome did, indeed, produce
fewer offspring than normal males, showing that half their
sperm had been sabotaged, and studies made with the aid
of electron microscopy showed that half the sperm produced by those males had distorted tails.

Just how the mutant chromosome does this trick isn't
known. But that doesn't detract from the significance of

the finding, which runs against the grain of evolution at the level of individual animals and species. Producing less offspring is clearly a disadvantage for males carrying this package of genes; but it is a good thing for the linkage group itself. It happens that in this case the linkage group can be easily identified because it carries both the cheating gene and the red-eye gene. But the same sort of thing may be going on, undetected, among chromosomes which carry genes that cheat but do not carry another mutation which shows up so clearly in the phenotype. And this is far from being the only case of cheating by genes during meiosis. One other example comes from a species much more closely related to us – another mammal, the house mouse. A fairly widespread genetic 'failure' in house mice produces individuals with abnormal tails and a variety of other defects. The mutations are highly damaging, and according to traditional understanding of Darwinian selection operating on the varieties produced by Mendelian segregation of genes and recombination they ought to be much rarer than they are. They are maintained because, as in the *Drosophila* example, mice carrying the linkage group that produces these mutations tend not to produce 'normal' sperm, even if they carry the normal linkage group on the appropriate homozygous chromosome.

Studies of this kind of mutation provide one approach for further, detailed investigation of recombination at work. They also raise the possibility of deliberately creating this kind of mutation in the laboratory – by breeding, or by genetic engineering – and using it to control insect pests. But in the present context, the most important point is that they highlight the mechanisms of evolution and natural selection.

Selection operates on individual *phenotypes*, not on individual genes. But reproduction very much involves individual genes and chromosomes.

Anthropomorphosising, we can say that a particular gene doesn't 'care' about the nature of the phenotype it inhabits, all it 'cares' about is reproducing itself. This is, therefore, also the driving force of the phenotype; only individuals that reproduce are an evolutionary success. As

Richard Dawkins, of the University of Oxford, has put it, we are all 'gene machines', with no purpose other than to copy genes and pass them on. The selfishness of genes, the struggle for survival among alleles of the same gene, between different linkage groups, and among chromosomes themselves, is no less real and no less important to an understanding of evolution than the struggle between individual members of a species, and between one species and another. And this understanding of the complexity of the processes operating within cells, especially those operating during meiosis, has, in the 1980s, put a whole new complexion on discoveries made half a century ago by one of the greatest biologists, Barbara McClintock.

# McClintock's maize

McClintock was born in 1902, two years after the rediscovery of Mendel's laws of heredity. She suffered from two handicaps in the scientific world of the early and middle twentieth century – first, she was a genius; secondly, she was a woman. She received full recognition only in the 1980s, with a flurry of belated awards, including the Nobel Prize in 1983, and an excellent biography, written by Evelyn Fox Keller, which appeared just before the announcement of the Nobel award. All of that derived principally from McClintock's second phase of work, in the 1940s and later, which showed that some genes control other genes, switching them on and off depending on circumstances, and that some genes can change their position on the chromosome, 'jumping' from place to place. None of this sounds too heretical in the context of the understanding of how mutations arise and genes cheat as outlined above. But it was regarded as distinctly odd, 40 years ago, for two reasons. First, the great inheritance from Mendelian genetics was the idea of the constancy of genes. Having only recently come over to accepting the idea of discrete 'particles' of heredity which followed well-defined laws, biologists weren't ready to swallow the idea that these constant entities could jump about within the chromosomes and manipulate each other. This was, remember, still more than a decade before even the beginning of Hiraizumi's study of cheating genes in *Drosophila*. But there was another, even more important reason why McClintock's work did not receive the recognition and acceptance it deserved at the time. In the 1940s, most biologists interested in determining the structure of genes were turning to physics and chemistry, focusing down on ever smaller components within the cell, and ultimately down to the genes and the molecules of life themselves. Many of this new breed of biologists were trained in the physical sciences and had no understanding of botany at all; even those who were trained in biology saw the study of whole plants as somewhat passé. McClintock was in a very small minority in continuing, from the 1940s to the 1980s, to seek fundamental genetic information by studying the whole organism. Not even a fast breeding organism like *Drosophila*, but a plant, maize, which produced only one generation each year. She looked like a throwback to Mendel's day; in fact, she resembled Mendel in quite another way, in being four decades ahead of her time with techniques and insights that her peers were simply not equipped to comprehend.

Maize had long been Barbara McClintock's chosen experimental organism. The plant, like Mendel's peas, is ideal in many ways. One obvious asset is the way variations are displayed, almost advertised, in the rows of corn seeds in the kernel. There is no need here for painstaking studies under the microscope, or catching eighth-inch-long flies and looking at their eyes; all you have to do is peel back the husk and see the patterns of different coloured seeds lined up before you. As McClintock expresses it, you get a 'feeling' for the whole organism – and that reveals a lot more about how genes work *together* than can ever be learned by studying individual molecules in the laboratory, even if those molecules are the molecules of life.

McClintock studied at Cornell University and obtained her PhD in botany in 1927, still not quite 25 years old. She decided to follow up the work on *Drosophila* which had led to the conclusion that linkage groups are carried on specific chromosomes, but working with maize instead of fruit flies.

This was a far more significant piece of work than is immediately apparent today, because in fact the work by Morgan and his colleagues had not *proved* the reality of linkage groups carried by chromosomes. It was all, strictly speaking, circumstantial evidence. Before McClintock began her studies with maize, many people had done breeding experiments, especially with *Drosophila*, and many other people had studied chromosomes. But 'they did not get together – they even worked in separate places'.* McClintock proposed to combine the two approaches, studying the chromosomes of the *same* plants that she would be using in the breeding experiments. And, much later, she told Keller of the reaction this proposal provoked: 'The people in genetics couldn't understand it. Not only that, they thought me a little mad for doing this.'†
With her first piece of post-doctoral research, McClintock was already moving herself away from mainstream ideas, demonstrating an unconventional, but effective, approach to the problems of the time.

She began the work alone, working slowly because of the long reproductive cycle of maize, but, for the same reason, having ample time to interpret the changes from one generation to the next, unlike the *Drosophila* specialists, sometimes almost snowed under with their new generations every two weeks. In 1929 she was joined by a young research student, Harriet Creighton, and together (but very much under McClintock's direction) they completed the experiments. The study confirmed and proved completely the reality of linkage groups, and went further than previous studies in identifying under the microscope precise changes in chromosomes associated with particular changes in the phenotype. In one example, the corn strain used was heterozygous for corn kernel colour, carrying chromosomes which produced either dark or light kernels. When the cells were stained and examined under the

* † Both quotations from Barbara McClintock are on page 45 of Keller's book.

microscope, it was found that the difference between coloured and colourless phenotypes was associated with a change in chromosome number 9, the chromosome in the coloured plants having a knob which was missing in those from the uncoloured plants.

The results were published in 1931, partly at the prompting of T. H. Morgan after a visit to Cornell where he learned of the work of Creighton and McClintock. Morgan knew, though he did not tell the Cornell team, that over in Europe Curt Stern was on the brink of completing similar studies of recombination of particular genes correlated with morphological changes in the phenotype of *Drosophila*. 'I thought', said Morgan afterwards, 'it was about time that corn got a chance to beat *Drosophila*.'* If Morgan's work in the previous decade had marked the beginning of the establishment of Mendelian genetics, McClintock's work with Creighton, and the confirmation from Stern's work, marked the end of the beginning. Morgan's Nobel Prize was awarded just two years after publication of the work by Creighton and McClintock which finally *proved* the reality of linkage groups and recombination; in view of the cautious conservatism of the Nobel Committee, always anxious not to give an award too soon, in case the recipient later proves to have made a mistake, this is probably no coincidence.

The same cautious conservatism surely prevented McClintock from receiving a Nobel Prize herself at about this time. Women just weren't honoured in that way in those days, at least not on their own. It might just be acceptable to give Marie Curie a Prize in the early 1900s, but only jointly with her husband Pierre (even though, with all respect to Pierre, he did not make the major contribution to their allegedly joint work). Had McClintock worked alone, perhaps she might have shared Morgan's Prize; looking back from 1984, it seems that the only explanation of why she didn't receive at least that much

---

* The quotation is taken from page 59 of *A Feeling For the Organism*, by Evelyn Fox Keller.

recognition was that her colleague, Harriet Creighton, was also a woman, and giving even half a Prize to two women together would have been quite beyond the pale in the eyes of the all-male Nobel Committee sitting in its smoke-filled room in Stockholm. The decision may not even have been conscious; quite possibly it never crossed the minds of the judges to take this slip of a girl seriously enough to consider her work on its merits. The irony is that by surviving to a ripe old age (Nobel Prizes are never awarded posthumously) McClintock eventually received the award for work that those Committee members of the 1930s would have found far more outrageous and incomprehensible even than the idea of a woman who was also a scientist.

The details of that story must come in their place. Although McClintock followed her own trail, which later rejoined the broad highway of biological progress, there was no doubt about the main challenge facing biology in the 1930s. Thanks to Morgan, McClintock, and the rest, the reality of genes had been made clear, and their location, on the chromosomes, had been identified. But what *were* genes? And *how* did they control the cell and the organism? These were questions that could not be answered by breeding experiments, nor even by observations under the microscope. The answers had to come from work at the molecular level – chemistry, not biology. Happily for the biologists, at just this time, in the early 1930s, chemists were equipped, for the first time ever, to tackle the problem of finding out just what molecules genes are made of. Chemistry was ready to come to the aid of biology, because a new understanding of physics, quantum physics, was in the process of transforming chemistry.

# PART TWO

# DNA

'It has not escaped our notice that the specific pairing we have postulated immediately suggests a possible copying mechanism for the genetic material.'

Francis Crick and James Watson
*Nature*, 1953

# DNA

It has not escaped our notice that the specific
pairing we have postulated immediately suggests a
possible copying mechanism for the genetic
material.

Francis Crick and James Watson
Nature, 1953

# CHAPTER FOUR

# QUANTUM PHYSICS

'Living things are composed of lifeless molecules. When these molecules are isolated and examined individually, they conform to all the physical and chemical laws that describe the behaviour of inanimate matter.'

The quotation comes from the first page of Chapter One of Albert Lehninger's excellent introductory text *Principles of Biochemistry*, and I cannot improve upon it. Living things are made of non-living molecules, and if we want to understand the chemistry of life we have to understand the chemistry of non-living molecules, and the laws of physics which underpin any understanding of chemistry. Terms like 'atom' and 'molecule' are part of the common language today, not reserved for the discussions of physicists and chemists, but we ought to refresh our memory of what the terms mean before moving on to more complex matters. All matter is composed of atoms, tiny fundamental constituents, originally thought to be indivisible, which form the building blocks of everything we see and feel. An element is a substance composed only of one type of atom – a diamond, for instance, is made up solely of billions upon billions of identical atoms, carbon atoms. Not all elements are so exotic or valuable as a diamond, however; iron is an element. But most things we come across in our everyday lives are not elements but compounds, combinations of two or more elements together. A compound is a substance made up of larger building blocks, called molecules; these are produced when atoms from the elements which make up the compound have combined. The gas carbon dioxide is a combination of carbon and oxygen. Each atom of carbon is combined with two atoms of oxygen, in this case, producing billions upon billions of essentially identical carbon dioxide molecules, denoted $CO_2$. Oxygen itself is present in the air we breathe, of course, but not as single atoms because the oxygen atoms themselves pair up to form di-atomic molecules, $O_2$.

## Atoms and molecules

The modern understanding of the atomic structure of matter goes back only to the late eighteenth century, when the French chemist Antoine Lavoisier investigated the way things burn. It was only during the second half of the nineteenth century that the idea was established of a gas being made up of a great number of little hard 'balls' (atoms or molecules) bouncing around and colliding with each other and with the walls of the container to produce an outward pressure. In 1897 electrons were discovered and identified as tiny negatively charged particles, little pieces that could be knocked off from atoms under the right circumstances; and in the twentieth century the picture that was established was of a cloud of electrons surrounding the dense central core of the atom, a core which was called the nucleus in conscious imitation of the name given to the central part of a living cell. Atomic nuclei themselves were found not to be indestructible, but to undergo both fission and fusion, the processes which power both the atomic bomb and nuclear reactors. For the story of chemistry, though, all that is needed is an understanding of how electrons behave, because the electron cloud is the visible face which an atom shows to the world, and it is the outermost electrons in that cloud that interact directly with other atoms when atoms join together to form molecules.

Before electrons were discovered, ideas about how elements combined, and why they should combine to form compounds always in particular ratios (such as one atom of carbon to each two of oxygen) could only be based on empirical rules. These were rules of thumb gleaned from many observations of many experiments, direct measurements of how much carbon dioxide, say, is produced when a certain amount of carbon is burnt in oxygen. Atoms seemed to form bonds with one another in accordance with certain rules, but just why those should be the rules of the game nobody knew. The rules said, for example, that one atom of carbon could combine with one di-atomic molecule of oxygen to make one molecule of carbon dioxide, a reaction written in shorthand as

$$C + O_2 \rightarrow CO_2$$

On the other hand, when hydrogen and oxygen combine it always takes two hydrogen atoms to 'satisfy' each oxygen atom:

$$2H_2 + O_2 \rightarrow 2H_2O$$

Or, in plain English, two di-atomic molecules of hydrogen combine with each di-atomic molecule of oxygen to make two molecules of water.

From this kind of evidence, chemists developed the idea that each kind of atom has the capacity to develop only a certain number of bonds with other atoms. Each hydrogen atom can form one, and only one, bond; each oxygen atom can form two bonds, and therefore has the capacity to latch on to two hydrogen atoms; and each carbon atom can form four bonds, and so latches on to two oxygen atoms. Sure enough, in line with this idea, four hydrogen atoms can combine with one carbon atom to form a molecule of methane:

$$2H_2 + C \rightarrow CH_4$$

The rule works very well indeed for a vast array of chemical reactions involving different elements and compounds, although carbon is itself something of a special case, as we shall see. But biological molecules, the molecules of life, are so much more complicated than simple molecules such as carbon dioxide, water, and methane, that in the nineteeth century there was still room to believe that something over and above the simple laws of chemistry was needed to explain their behaviour. After all, at that time nobody knew what the laws of chemistry really were, so there was no way of checking to see if they could explain the behaviour of very large, complex molecules.

# Molecules of life

Living organisms are indeed complicated and highly organised, but it is now clear that no new scientific laws are required to explain their complexity and organisation. Just as living organisms like ourselves are composed of many different kinds of organ, each with its own specific part to play in ensuring the proper functioning of the whole organism, so at a deeper level the tissues of all living things are composed of many kinds of complex molecules. Each of those chemical compounds, like each organ in the body, seems to have a specific role to play in the healthy functioning of the whole, and some of those molecules may contain tens of thousands, even hundreds of thousands, of atoms.

The essence of life – ultimately due to the behaviour of those complex molecules working together – is the ability of living things to extract energy from their environment and to use that energy both to build up their own complex structures and to copy themselves – to reproduce. For all life on the surface of the Earth, the ultimate source of that energy is sunlight, trapped in plants by photosynthesis. Cells function as chemical engines, storing and transmitting energy in chemical form, as energetic molecules which can give it up when and where required; animals tap that storehouse of energy by eating plants, or by eating other animals which have themselves eaten plants. (Very recently, organisms have been found in the deep ocean, far from any source of sunlight; they derive their energy from underwater volcanic hotspots and lead their lives quite cut off from the ecological web above them. Like us, however, they depend on an outside source of energy.)

Most of the chemical compounds in living things are carbon compounds, and this is why the study of complex carbon compounds (anything a little more complex than carbon dioxide) is called organic chemistry. Typically, the molecules that are important for life also include atoms of hydrogen, oxygen, phosphorus and nitrogen. The key family of compounds involved in both the structure and functioning of living things is the proteins; their very name means 'foremost' (among organic molecules) and a single cell of the common bacterium *Escherichia coli* (which lives, among other places, in your gut) contains roughly 5 000 different kinds of organic molecule, with 3 000 of those varieties different types of protein. In your own body, there are more than 50 000 different kinds of protein, contributing their part to the healthy operation of the whole organism. There is such a variety of proteins, however, that it is unlikely that any of the proteins in your body is exactly the same as any of the proteins found in *E. coli*; each species has its own set of proteins. At the next structural level down, though, all living things are very much more alike.

The amount of material present in a molecule is indicated by its molecular weight, on a scale in which the weight of one atom of hydrogen (the lightest element) is 1. On this scale, one carbon atom weighs 12 units. Molecular weights of proteins range from a few thousand up to several million. But all these proteins are made up of much smaller sub-units, called amino acids, linked together in chains which fold up upon themselves. The molecular weights of amino acids are typically a little more than 100 units on the same scale, which gives some indication of just how many amino acid groups there are in the larger proteins. But the really important point is that all proteins, in all living things on Earth, are composed of different permutations and combinations of the *same* basic amino acids, just 20 different varieties. On their own, amino acids have no intrinsic biological properties – they are not living molecules. Joined together to make proteins, however, they become the very stuff of life. There are, indeed, more amino acids than the 20 used in the manufacture of proteins; the fact that all living things on our planet use the same 20 building blocks of life, in much the same way, is one of the strongest hints that we are all descended from some single original molecule that learnt the trick of life, the trick of extracting energy from the environment and using it to make copies of itself. Other molecules may, perhaps, have learnt the same trick at more or less the same time. But only one variety, it seems, won out in the struggle for survival and has left descendants on Earth today.

The number 20 is conveniently close to a number familiar in everyday life that immediately makes clear the enormous potential for variety among proteins. Proteins are made up of combinations of 20 different amino acids, strung out in different orders along chains which may be as short or as long as you please. Books are made up of combinations of 26 letters (in the English alphabet), plus a few punctuation marks, strung out in different orders along chains which may be as long or as short as you please, and are conveniently 'folded' to lie along the pages of the book. Just as the letters of the English language can be arranged into an enormous variety of books, so the 20 amino acids used by life can be arranged into an enormous variety of different proteins. The proteins are manufactured by cells, and the ultimate role of the genetic material carried by the chromosomes is to instruct the cell how to manufacture different kinds of protein, and when. Clearly, the information stored in the chromosomes must be comparably detailed and complex as the information 'coded' by the 20-character amino acid 'alphabet'. In proteins, very large molecules – macromolecules – constitute the message, and relatively small molecules – amino acids – form the letters of the code. A very similar coding system carries the genetic message in the chromosomes, in the form of the famous double helix. The story of the cracking of that genetic code forms the bulk of the rest of this book. It follows directly from the understanding of the chemical bond developed by Nobel laureate Linus Pauling in the late 1920s and early 1930s, an understanding which in turn depended on the developments in atomic physics during the first quarter of the twentieth century – an understanding of the nature of the electron and its behaviour, within the framework of quantum physics, as part of the atom.

# Physics transformed

At the beginning of the 1890s, the world of the physicists seemed to be orderly and reasonably well understood. That world divided into two parts, matter and radiation. Matter could be regarded as made up of atoms – tiny, hard objects which were indestructible and bounced off each other like pool balls colliding, but which could also stick together, in some poorly understood way, to make molecules. Electromagnetic radiation, most notably light, could be explained even more completely, thanks to the equations developed by the Scot James Clerk Maxwell, as a form of vibration, ripples in the 'luminiferous ether' similar to the ripples on a pond. But as the end of the century approached, the orderly appearance of the physicists' world was disrupted by a series of new discoveries. Physics, chemistry and biology would never be the same again.

In November 1895 the German physicist Wilhelm Röntgen discovered a previously unknown form of radiation, now called X-rays, which passed right through most substances, could affect photographic plates to produce images even though the plates were securely wrapped in paper, and had many other curious properties. Within months of the discovery, even without an understanding of the nature of this new kind of penetrating radiation, X-rays were being used in medical diagnosis, providing images of broken bones and internal organs for doctors without any need for surgery to open the body up. Röntgen's discovery, breaking the mould of nineteenth-century physics, makes a convenient marker for the beginning of modern physics; indeed, in 1901 he received the first ever Nobel Prize in physics for this work. But the discovery was to a large extent fortuitous, and, strictly speaking, it came out of turn in the unfolding new picture of the world that was emerging at the end of the nineteenth century.* As long ago as 1858

---

* The outline presented here of how physics was transformed from the 1890s to the 1920s can only be a quick sketch, setting the scene for the new understanding of atoms which transformed chemistry in the 1930s. The full story is told in my book *In Search of Schrödinger's Cat*.

physicists had found another form of radiation, called cathode rays. This radiation is produced when two metal plates in a glass tube are connected in an electrical circuit so that one becomes positively charged (the anode) and the other negatively charged (the cathode). When air is pumped out of the tube and only a trace of gas is left behind, cathode rays are emitted from the cathode. They can make the glass wall of the container fluoresce when they strike it, and they can make the trace of gas left in the tube glow brightly with a characteristic colour which depends on the composition of the gas. This is the basis of the neon tube now so familiar in advertising, and of a common form of strip lighting. Röntgen's discovery of X-rays came from his investigations of cathode rays – he found that X-rays could be produced when the cathode rays struck certain materials. The cathode rays themselves had been tentatively identified as charged particles, carrying negative electricity, by Sir William Crookes, in 1879. In 1897 another Englishman, J. J. Thomson, proved that this was the case, and measured the ratio of the charge $(e)$ to the mass $(m)$ of the particle we now call the electron.

The very small size of this ratio, $e/m$, and the fact that it was exactly the same for all 'cathode rays' led Thomson to the conclusion that the rays are actually a stream of identical particles, that these particles are part of every atom, and that the electrons from every atom are identical. He had proved that the atom could be subdivided, and that it could no longer be regarded as an indestructible billiard ball. Thomson also took the first step towards explaining another phenomenon, discovered in 1888. This is the photoelectric effect. When light shines on to a metal plate, negatively charged particles are emitted from its surface, provided that the wavelength of the light is shorter than some critical value, which depends on the metal being used. (The optical spectrum of visible light – the rainbow – extends from red, which has the longest wavelength, to violet, which has the shortest. The electromagnetic spectrum continues, invisible to our eyes, at both ends, into the long wavelength infrared and the short wavelength ultraviolet, and beyond.) Thomson identified the negatively charged

particles produced in this way as electrons, and set the
scene, as we shall see, for a further step forward in physics.

In between the discovery of X-rays and Thomson's definitive identification of cathode rays as electrons, in 1896
the French physicist Henri Becquerel discovered that the
element uranium spontaneously emits another form of
radiation. This behaviour was given the name radioac-
tivity. By 1900, the French husband and wife team Pierre
and Marie Curie had isolated other radioactive elements,
and the British physicist Ernest Rutherford had identified
three different kinds of radiation produced by these elements, dubbing them alpha, beta and gamma radiation in
order of their penetrating ability, with gamma rays being
the strongest. So, at the beginning of the twentieth century
physics was in the throes of a transformation. The atom
was *not* indestructible; new forms of radiation had been
discovered and had yet to be explained; and there was a
new particle to be investigated, the electron, which seemed
to be a chip off the atom. For the crucial developments in
chemistry which paved the way for the science of molecular
biology, the story of the electron and its place in the atom is
all-important. But that story can only be understood in the
context of the new understanding of both matter and radiation which emerged by 1926.

# Particles and waves

Alpha radiation doesn't really come into this story directly.
In the early twentieth century this form of radiation was
identified as fast moving particles each exactly equivalent
to an atom of helium from which two electrons had been
removed – a positively charged helium ion. The beta radiation was also quickly identified in terms of fast moving
particles – electrons, or cathode rays. But gamma radiation, like X-radiation, continued to pose puzzles for some
time. After a period of initial confusion, by about 1905 it
was widely accepted that gamma rays are a more intense
form of X-radiation, more powerful and even more penetrating. Explain X-rays and, by implication, you would

explain the nature of gamma rays. But what were X-rays? Up to 1912, most physicists assumed that X-rays were short but intense bursts of electromagnetic radiation (pulses) propagating in a similar, but not identical, way to light. The radiation only appeared to be continuous, it was argued, because the pulses followed so rapidly one after another, in much the same way that the distinct separate rings of an electric bell blend into a continous buzz if the speed of the bell hammer increases. The main reason for this assumption seems to have been that X-rays (and gamma rays) could affect photographic plates. Previously, the only thing known to affect the photographic emulsion in this way was light, so if X-rays affected photographic emulsion then, surely, they must be a form of intense light. But some physicists, notably in England, disagreed. Following the success of Thomson's identification of cathode rays as charged particles, British physicists tried to explain the other newly discovered forms of radiation in a similar way. William Henry Bragg, in particular, endeavoured to explain the behaviour of X-rays in terms of a stream of particles carrying no electric charge.

While the debate about whether X-rays were waves or particles continued in the first decade of the twentieth century,* it at least seemed clear to most scientists that electrons were particles and ordinary light was a wave phenomenon. Even though the German physicist Max Planck had, as early as 1900, invoked the idea of distinct units of electromagnetic energy to explain the detailed nature of the spectrum of radiation produced by a hot object, this was widely taken as meaning that atoms could only accept or emit light in distinct units, not that light only existed in distinct units. Possibly the only person in the early 1900s who seriously entertained the idea that light might also behave like a stream of particles was Albert Einstein, then just starting out on his scientific career and

* The X-ray story, and its relation to the wave/particle puzzle, has recently been told in fascinating detail by Bruce Wheaton, in *The Tiger and the Shark*.

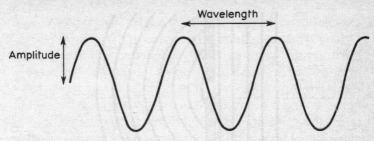

Figure 4.1 A wave.

largely unknown. In 1905, Einstein proposed an explanation of the photoelectric effect which depended on the assumption that light came in distinct packets, as particles – what would now be called photons. Provided it had enough energy, a single photon could knock an electron out of a metal atom, and on this picture more energy exactly corresponds to shorter wavelength on the old picture. Einstein's interpretation of the evidence explained, in a way that the wave theory could not, all of the observed details of the photoelectric effect.

One person who took Einstein's proposal seriously was the American experimental physicist Robert Millikan. But he didn't accept Einstein's interpretation of the evidence, and his approach to the puzzle started out as a serious attempt to *dis*prove Einstein's idea that light might come in the form of particles – that it might be 'quantised'. Starting out in 1906, it took Millikan and his collaborators until 1914 to prove to his own satisfaction that Einstein was right, that the photoelectric effect could best be explained in the way Einstein had suggested in 1905, and that for some purposes the best description of light is indeed as a stream of particles. Photons seemed to be real, although not perhaps quite as real as electrons, and Einstein was duly awarded a Nobel Prize for this insight, in 1922 (it was actually the 1921 prize, held over for a year). But while physicists were being reluctantly forced to the conclusion that light displayed properties of both wave and particle, depending on the kind of experiment it was involved with, the debate about the nature of X-radiation seemed to be being settled in favour of the wave idea.

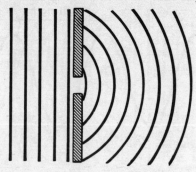

Figure 4.2 Parallel, or plane, waves passing through a hole spread out in a semicircular pattern.

The initial breakthrough came in Germany, from the work of Max Laue. One of the most striking features of waves is the way they can interact with each other, interfering to produce distinctive patterns of peaks and troughs. This is true for ripples on the surface of a pond, and it is also true for two beams of light, carefully prepared to be in step with one another by letting light from a single source – a lamp – pass through two pinholes in a sheet of cardboard. Light spreading out from each of the pinholes overlaps and interferes to produce a characteristic pattern of bright stripes and dark shadows – a diffraction pattern – on a second card held up on the other side of the pinholes from the lamp. This, and similar, effects can only be explained in terms of wave motion, and such experiments had been the basis of the wave interpretation of light in the nineteenth century.

By 1912, it was clear that if X-rays were to be interpreted as waves anything like light waves then they must have very short wavelengths, corresponding to very high energy. Because of this, the only way to produce a diffraction pattern from a beam of X-rays is to use the equivalent of two or more 'pinholes' placed very close together indeed. Laue realised that the spacing between atoms in a crystal would be just about right to do the trick, and at his suggestion Walther Friedrich and Paul Knipping, of the Institute of Theoretical Physics in Munich, carried out the experiment. They directed a narrow beam of X-rays at a single

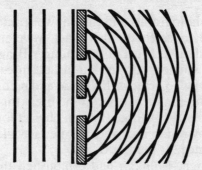

Figure 4.3 The circular waves from two such holes interfere with one another. In some places they add together, in others they cancel out.

crystal of zinc sulphide, and placed a photographic plate on the other side of the crystal. When the plate was developed, it showed the pattern that ought to have been produced if the X-rays were waves and interfered with each other like waves of light. But at that time it wasn't clear exactly how the interference was occurring. A crystal consists of an array of atoms in a lattice, evenly spaced and stretching out in uniform layers in all directions, and this is much more complicated than the situation for two pinholes in a sheet of cardboard. The photographs looked like complicated diffraction patterns, but the full explanation was still to come.

It wasn't long in coming. News of the discovery quickly spread to Bragg in England, and to his son, William Lawrence Bragg, then a student at Cambridge University. Although the older Bragg (William) tried at first to explain the phenomenon in terms of X-rays as neutral particles, his son (usually referred to as Lawrence) soon realised that the Germans had indeed found an X-ray equivalent to optical diffraction. It was Lawrence Bragg who worked out the rules which made it possible to predict exactly where the bright spots would be produced when an X-ray beam with a particular wavelength struck a crystal lattice with a particular spacing between atoms at a particular angle. He produced an equation called Bragg's Law from which it is possible to work backwards, using the measured bright spots on the photographs and the known spacing of atoms in the crystal to calculate the exact wavelength of the original X-radiation. Most of the younger Bragg's work at this time was carried out and published in collaboration with his father, who acted as his guide and mentor but who was, in fact, the junior partner in the breakthrough work. They shared the 1915 Nobel Prize in physics for their joint research – Lawrence received the news while on active service in France, at the age of 25.

By 1915, then, Millikan had established that Einstein was right in suggesting that light could behave like a particle, as well as like a wave, and almost exactly at the same time the puzzle of whether X-rays were particles or waves had been resolved with proof that X-rays were, like light, waves. The only snag being, of course, that if X-rays were like light and if Einstein was correct then X-rays, also, should sometimes behave like particles. William Bragg had, indeed, talked of possibilities along these lines in 1912, suggesting that a beam of X-rays, and even a beam of light, might be considered as made up of a wave-like component and a particle-like component existing side by side. But the suggestion made no great impact at the time. However, experiments showed that X-rays, like light, could produce the photoelectric effect, knocking electrons out of metals in the way Einstein explained in terms of photons. And in the early 1920s a definitive series of experiments by the American Arthur Holly Compton showed that X-rays 'must' be particles.

The experiments involve measurements of the way the energy, or wavelength, of the X-ray beam changes when the radiation is scattered by being bounced off atoms. Without going into details, both the energy and the direction of the scattered beam depend on the way the X-rays and the atoms interact, and these measurable properties are different for waves which are scattered like light waves than they are for hard little balls which are bouncing off the atoms. The measured change in energy of the X-rays in this process, conventionally interpreted as a change in wavelength, could only be explained in terms of particle collisions, little balls bouncing off atoms, or off electrons.

Both Compton and Peter Debye, a Dutch scientist worl
in Zürich, independently arrived at the same conclusion in
1922.

So by 1923 the physicists' view of the world looked very
different from the view thirty years before. The distinction
between particles and waves had become blurred.
Although it had been satisfactorily established that X- and
gamma rays were indeed more energetic versions of light,
it had also been established that all three forms of radiation
sometimes behaved like waves and sometimes behaved
like particles. It all depended on what sort of experiment
you chose to carry out to test their nature, and there was no
simple answer to the question of whether light, or X-radiation, or gamma rays were 'really' a wave phenomenon or
another aspect of the particle world. The scene was set for
the next bombshell, one of those delightful insights that
look so obvious with hindsight but came like a bolt from the
blue at the time. If waves can sometimes behave like
particles, why can't particles sometimes behave like
waves?

## Electron waves

Louis de Broglie, born in 1892, was the second son of a
French nobleman, the Duc de Broglie. His brother
Maurice, seventeen years older than Louis, became a
highly respected physicist, closely involved in the investigation of the mysterious X- and gamma rays in the years
up to the First World War. Maurice had inherited the
family title in 1906, as Louis was to inherit it from Maurice,
and carried out research in a private laboratory in his Paris
residence. Although educated at the College de France, as
a gentleman amateur he hardly fitted the conventional
image of a scientific researcher of the first rank, even in
those days, and his position was even more unusual
because he worked in an area where France had no great
tradition, and where the major contributions were coming
from Britain and Germany. But he successfully presented
his doctoral thesis in 1908, and, of course, the security of

his background left him free to investigate what he liked, with no need to fit in with the French scientific establishment. He rapidly made a name for himself, pioneering techniques used by researchers such as Rutherford in experiments with X-ray diffraction from crystals.

Hardly surprisingly, young Louis was strongly influenced by his brother. Fourteen when their father died, although the family originally intended him for a career in the diplomatic service (his first degree was in history!) Louis turned increasingly to physics, where his brother's work gave him an opportunity to see the dramatic developments of the time at first hand. Maurice was, for example, the scientific secretary of a major conference on the new physics held in 1911, the First Solvay Congress, and gave Louis a chance to study the material which he was preparing for publication. In later years, Louis recalled that this was the event that made him determined to investigate the nature of the mysterious quanta which Max Planck had introduced into theoretical physics. But the ambition was delayed by the war, during which Louis served as a radio telegraph operator, based on the Eiffel Tower.

Afterwards, he worked closely with Maurice on various studies of X-rays. In April 1921, Maurice reported to the Third Solvay Congress experiments which provided the first clear indication of the photoelectric effect associated with X-rays; soon afterwards, one of Rutherford's students in Cambridge, Charles Ellis, confirmed the results and extended them to gamma rays. The award of Einstein's Nobel Prize followed closely on the heels of this confirmation by de Broglie and Ellis of the validity of his description of the photoelectric effect for other forms of radiation than light.

Maurice's experiments showed clearly that during the photoelectric process an X-ray gives up the whole of its quantum of energy – there is no combination of wave and particle moving, as it were, side by side. It wasn't that the X-ray was a combination of a wave *and* a particle, but rather that the X-ray possessed some mixture of the properties which are associated, in the everyday world, with particles and waves separately. Maurice's work, discussed extensively with his brother, set Louis thinking along new lines. First, he read up all of Einstein's papers on the nature of light. Einstein's most famous equation is, of course, the one which relates the energy ($E$) locked up in a mass ($m$) and the speed of light ($c$):

$$E = mc^2$$

The particles of light – photons – required by the photoelectric effect do not have any mass in the everyday sense, but they do carry energy, and that means that they also carry momentum, a property associated in the everyday world with moving objects that do have mass. Momentum is a measure of how much push it would require to stop the moving object, or how much it takes to give it the velocity ($v$) in the first place. It is usually written

$$p = mv$$

Planck's equation, which indicates the amount of energy contained in a quantum of radiation – a photon, in the new terminology – introduces a new constant, called Planck's constant and written as $h$. Instead of dealing in terms of wavelength, it is usually written in terms of the frequency of the radiation, which is just 1 divided by the wavelength, and is denoted by the Greek letter nu, $v$. The equation is

$$E = hv$$

Einstein pointed out that each photon must carry a momentum given simply by dividing its energy by its speed $E/c$. Putting this in terms of Planck's equation, that meant

$$p = hv/c$$

Frequency and wavelength are related by a simple formula. For a wave of frequency $v$ moving at speed $c$, the wavelength, denoted by the Greek letter lambda, $\lambda$, is simply $c/v$ So this equation can also be written

$$p = h/\lambda$$

The equation Einstein deduced in 1916 related momentum, a property which is associated with particles, to frequency, or wavelength, a property associated with waves.

As the work of Maurice de Broglie and others developed in the early 1920s, the link was taken more and more seriously, but only for radiation like light, electromagnetic radiation, including X- and gamma rays. It was Louis de Broglie who not only appreciated that the equation must work both ways, but also worked out a self-consistent interpretation of what that would mean for fundamental particles, notably electrons, and atoms. In a series of papers published in 1923, which formed the basis of his doctoral thesis presented late in 1924, de Broglie argued that the frequency, or wavelength, associated with an electron carrying momentum $p$ must be every bit as real as the momentum associated with a photon of energy $h\nu$. One of de Broglie's examiners, Paul Langevin, sent a copy of the thesis to Einstein; it was Einstein who publicised the ideas in it, introducing them to the wider scientific world with his stamp of approval, and ensuring that, outrageous though they seemed, they were almost immediately taken seriously.

De Broglie's equation can be written in terms of wavelength, $\lambda$:

$$\lambda = h/mv$$

or

$$p\lambda = h$$

In other words, every particle or object with a mass $m$ and velocity $v$ has associated with it a wavelength $\lambda$. The bigger the mass of the particle (strictly speaking, the bigger its momentum, $mv$), the smaller its wavelength. Planck's constant itself is very, very small. It checks in at a value, determined from studies of the radiation from hot objects, of $6.63 \times 10^{-27}$ erg seconds – that is, a decimal point followed by 26 zeroes and 663. The only way to end up with a measurable wavelength out of de Broglie's equation is to have a mass which is comparably small; the smaller the mass, the bigger the wavelength of the particle. That is why the phenomenon of the dual nature of particles and waves had never been noticed in the everyday world. But the mass of an electron is a little over $9 \times 10^{-28}$ grams, a decimal point followed by 27 zeroes and a 9. According to de Broglie's equation, electrons should have a measurable wavelength.

In 1927, confirmation of the reality of Louis de Broglie's 'matter waves' came from both the United States and Britain. Clinton Davisson and Lester Germer, working at the Bell Laboratories, and George Thomson and Alexander Reid, on the other side of the Atlantic, independently found the electron equivalent of the now standard X-ray diffraction patterns. Davisson and Thomson shared the Nobel Prize in physics in 1937; de Broglie had already received the award in 1929. Together, they had established that what we think of as two separate phenomena, particles and waves, are simply opposite faces of the same coin. In the everyday world, the wave nature of matter is undetectable because the masses involved are so large compared with Planck's constant, $h$. But the smaller the mass of the particle involved, the more important its wave nature becomes, until for the massless particles of electromagnetic radiation – the photons – both facets of nature are equally important.*

It is no use asking whether a photon, an X-ray or an electron is 'really' a particle or a wave. The names 'photon', 'X-ray' and 'electron' are simply labels which we attach to certain natural phenomena. When we make certain measurements of those phenomena – perform certain experiments – the results we get can be interpreted, for convenience, in terms of the behaviour of particles in the everyday world. When we make other tests, the results we get are most conveniently interpreted in terms of the laws of physics that describe waves, such as ripples on a pond. The answers we get from nature depend not just on the questions we ask, but on the *kind* of questions we ask. Ask particle questions and we get particle answers; ask wave questions and we get wave answers. But the natural phenomenon itself is not 'really' either a wave or a particle. It is

---

* For massless particles, of course, it is essential to work in terms of momentum, not mass, since otherwise all the terms in the equations become either zero or infinite.

something we have no everyday experience of at all, something which is sometimes called a 'wavicle'.

The importance of all this for biology is that an understanding of the wave nature of the electron is a fundamental prerequisite for an understanding of the structure of atoms and the way atoms join together to form molecules. Chemistry depends on quantum physics; biology depends on chemistry. The time has very nearly come to return to the story of the development of chemistry and biology in the 1930s. But first, there is one more strange phenomenon to take on board. The idea of wave/particle duality may seem strange enough; but it becomes almost commonplace compared with the quantum mechanical phenomenon called uncertainty.

# Uncertainty

In quantum physics, uncertainty is a definite thing. It can be measured, very precisely, and is governed by equations and laws, like other physical phenomena. The history of how this concept became an integral part of physics is even more complicated and tangled than the rest of the story of how quantum physics emerged during the 1920s, but the physical ideas lying behind what has become known as the uncertainty principle can be stated fairly simply in terms of the modern understanding of particle/wave duality.*

It was a German physicist, Werner Heisenberg, who came up with the idea in 1927. In the everyday world, a wave is a spread out thing. The ripples on a pond stretch out over a long distance, and it is hard to be sure exactly where the string of ripples – the wave train – begins and ends. But a particle is a very well-defined thing, which occupies a definite place at a definite time. How can these two conflicting images be reconciled, as they must be if an

---

* The approach I use here is now standard; I have borrowed it in particular from Fritjof Capra's *The Tao of Physics* (Bantam edition, 1977, Chapter 11).

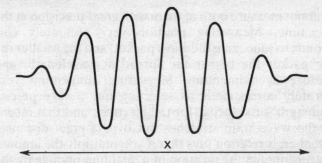

X

Figure 4.4 A particle can be described in terms of a wave packet, like this one. The packet is spread over a distance X; this represents the uncertainty in the location of the particle.

electron is to be regarded as both wave and particle at the same time? The appropriate image is of a little package of waves, a short wave train which only extends over a small distance, a distance roughly corresponding to the size of the equivalent particle. There is no difficulty about constructing such wave packets, as they are called, in the real world. The mathematics describing such phenomena are very well known. But the only way to create a wave packet which is localised in space is to allow waves of different wavelength to interfere with one another. And the smaller the wave packet the more variety of different wavelengths is needed to keep it tightly confined. The spread in wavelength depends only on the size of the wave packet, and has nothing to do with quantum effects – it happens in exactly the same way for waves in the everyday world, ripples on a pond and the like. But the spread in wavelength does have implications in terms of quantum physics, because we now know that a spread in wavelength must correspond to a spread of momentum over an equivalent amount. At the same time, even though the wave packet may be very small, it always has some physical spread in space itself.

So Heisenberg deduced two things: for a small particle like an electron, it is impossible to measure its position precisely, and it is impossible to measure its momentum precisely. In theory, we can measure either property as accurately as we like, just short of absolute precision. But

we cannot measure both of them with great precision at the
same time. Measuring position very accurately corresponds to squeezing the wave packet, and the smaller the
wave packet the bigger the spread in wavelength, and
therefore in momentum. Measuring momentum very
accurately corresponds to selecting out a very precise
wavelength, or velocity, for the electron, and that means
that the wave train stretches out over a great distance.
Heisenberg's relation says that if you multiply the amount
of uncertainty in the position of a quantum particle by the
amount of uncertainty in its momentum then the product
can never be less than Planck's constant divided by $2\pi$.
This is *not* simply a practical limit, an indication that our
measuring techniques are imperfect. It is a fundamental
law of nature, which says that there is no such thing as a
particle which has a precisely defined position and a precisely defined momentum at the same time. We cannot
know simultaneously exactly where a particle is *and* exactly
where it is going. The law is very closely linked tu the dual
wave/particle nature of things, but it is telling us something
at once both more subtle and more profound.

Notice the way I said that by measuring momentum
precisely we are *selecting* a wavelength for the electron. It
is no longer merely the case that the answers we get from
nature depend on the questions we ask. What the uncertainty relation is telling us is that what nature *is* depends on
the questions we ask. By choosing to measure the momentum of an electron, or beam of electrons, very accurately,
we are creating uncertainty in the position of the electrons;.
by measuring the position of an electron very precisely, our
experiment itself produces uncertainty about the wavelength, or momentum, of the electron. This is just the edge
of the very deepest and most mysterious of quantum mechanical pools, a glimpse of the way in which the experimenter, or observer, becomes part of what he is observing. In
fact, it makes more sense to say that *neither* the position nor
the momentum of an electron has any meaning at all until
one of them is measured. And this uncertainty extends to
other pairs of properties, such as the time a fundamental
event takes and the amount of energy involved in the

event. These are the mysteries I discuss in my book *In Search of Schrödinger's Cat*; for now, the time has come to draw back from the brink of such deep water (perhaps with a sigh of relief) and return to the really quite sensible behaviour, as far as particle/wave duality and uncertainty will allow, of individual atoms.

# Atoms

Atoms and molecules – even complex molecules – are almost unimaginably small by human standards. But even individual atoms are very large on the quantum scale. One way to get this in perspective is to calculate the wavelength that corresponds to a typical atom or molecule. The molecule we are going to be most interested in later is DNA, and the wavelength of a DNA molecule in the chromosomes in your body is $10^{-14}$ of a metre – a decimal point followed by thirteen zeroes and a 1. At room temperature, a typical atom of oxygen in the air that you breathe has a de Broglie wavelengh of about $4 \times 10^{-11}$ m. The wave properties of whole atoms and molecules can be completely ignored for almost all practical purposes. But atoms are partly made of electrons, and the wave properties of electrons cannot be ignored. The electron is, effectively, the largest particle for which wave-particle duality has to be taken into practical account, and that is what defines the behaviour of atoms and decides how they join together to make molecules.

Our present understanding of atoms is entirely a twentieth-century concept. It was only in 1911 that Ernest Rutherford demonstrated that atoms must be largely insubstantial clouds, with most of their mass concentrated into very tiny, dense cores, called nuclei because of the superficial similarity to the nuclei of cells. Over the next ten years or so, Niels Bohr, the great Danish physicist, was largely responsible for developing the first reasonably satisfactory theory of how quantum physics could be combined with Rutherford's picture of the atom to produce a theoretical model involving electrons in a cloud around the nucleus, making up most of the volume, but scarcely any of the mass, of the atom. And it took the developments in the second half of the 1920s, following the realisation of the wave nature of the electron and the development by Erwin Schrödinger of a wave equation to describe the behaviour of an electron, to produce a really satisfactory model of how it all works. I have dealt at length with this unfolding story in my book *In Search of Schrödinger's Cat*; it is so difficult to do justice to the convoluted tale of the historical development of the quantum theory of the atom in a limited space that I shall make no attempt to cover the same ground here, but will simply give the outlines of the picture of the atom that emerged by the end of the 1920s and remains our best model to this day.*

We now know that atomic nuclei are composed of two kinds of particles, protons and neutrons, squeezed tightly up against one another. (Protons and neutrons are themselves composed of another kind of particle, the quarks, but details at that level have no bearing on our tale.) The proton has an electric charge which is positive in sign, so that protons are attracted towards the negative electrode in a circuit, and has exactly the same size as the charge on the electron, which is negative. The mass of a single proton is $1.672 \times 10^{-27}$ of a kilogram, such a ludicrously small mass that it makes more sense to define the proton mass as 1 unit and measure atomic masses on that scale; more informatively, the proton's mass is 1836 times the mass of an electron. The other nuclear particle, the neutron, weighs in at 1839 times the mass of an electron, and it has no electric charge. Together, protons and neutrons are collectively referred to as nucleons; the most common form of carbon has twelve nucleons in its nucleus, and the atomic mass unit, or dalton, is officially defined as one-twelfth of the mass of a carbon nucleus. For our purposes, however, we can think of both the proton and the neutron as having a mass of one dalton.

---

* Apart from my own book, written from the viewpoint of a physicist, *Chemistry*, by Linus Pauling and Peter Pauling, gives a good account of the way the structure of the atom was determined, and summaries can be found in many popularisations of science.

Atoms are electrically neutral. Every proton in the nucleus is balanced, electrically, by the presence of one electron in the surrounding cloud. So the simplest atom consists of one proton and one electron. This is the structure of an atom of hydrogen. Because the electron cloud – in this case, the single electron – provides the visible face of the atom to other atoms, and the means by which atoms interact, the number of electrons in the cloud is what decides the atom's chemical properties. In other words, the number of electrons decides what chemical element the atom is. This is the same as the number of protons, and it is more common to find the statement that it is the number of protons in the nucleus that decides which element an atom belongs to. But the two statements are equivalent, and we are especially interested in electrons. Adding a neutron to the nucleus makes the atom heavier, but has very little effect on its chemical properties. So the next most complicated atomic system, with one proton and one neutron in the nucleus, and one electron outside, is still hydrogen – called, for obvious reasons, 'heavy hydrogen', or deuterium. It is only when another proton is added to the nucleus, and another electron outside, that we get a different element, in this case helium. The most common form of helium actually has two protons and two neutrons in the nucleus, plus the two electrons outside.

So it goes on. Mentally adding protons and neutrons to the nucleus and electrons to the surrounding cloud we can build up a picture of atoms corresponding to different elements. There are usually at least as many neutrons as there are protons, except in the lightest atoms, and many elements exist in stable forms with different numbers of neutrons, forms called isotopes. An oxygen atom, for example, contains 8 neutrons and 8 protons, plus 8 electrons, and has a mass of about 16 daltons; one type of uranium atom contains 92 protons and 146 neutrons, is surrounded by a cloud of 92 electrons, and has a mass of 238 daltons, 238 times the mass of a proton, or of a hydrogen atom. This is one of the biggest stable atoms.

The physical size of a nucleus depends on how many nucleons are clustered within it. The radius of a proton is a little more than one-thousandth of a millionth of a milli-metre, about $1.2 \times 10^{-15}$m. For our purposes, we can think of nuclei as being made of little billiard ball particles of this size jammed together. Even 238 of them don't take up very much room. The electron cloud around the nucleus is a much fuzzier thing, and it has no sharply defined edge, but in round terms most atoms are more or less the same size (the bigger ones are only a few times bigger than the smaller ones) and a typical atomic radius is about $10^{-10}$ m. In other words, the size of the atom is $10^5$ times – 100 000 times – bigger than the size of the nucleus, in terms of radius. Because volume goes as radius cubed, another way of looking at it is to say that the volume of the atom is $10^{15}$ times the volume of the nucleus. No wonder the elec-tron cloud dominates the observable properties of the atom and its chemical behaviour. But what is the structure of the cloud, and since opposite electric charges, like opposite magnetic poles, attract one another, why don't all the electrons fall into the nucleus and stick like glue to the protons?

# Electrons and atoms

It took a combination of the work of Niels Bohr, Louis de Broglie, Schrödinger and several others to establish what goes on in the electron cloud. Their work built on chemical evidence that goes back to the 1860s and the work of the Russian chemist Dmitri Mendelyeev, who was the first person to develop a periodic table of the elements, in which different elements that had similar chemical proper-ties were arranged together in columns. The concept was refined and developed further by Mendelyeev and others in the 1870s; those pioneers catalogued the different ele-ments in terms of atomic mass, but since we now know that chemical properties depend on the number and arrange-ment of electrons in the cloud around an atomic nucleus, it should be obvious, with hindsight, that these catalogues of similarities between elements are, in fact, picking out ele-ments which have similar electron clouds. The mass of an atom depends on the number of protons in the nucleus, as well as the number of neutrons, and for most elements the presence of different isotopes does not distort the picture too much.

Of course, the nineteenth-century chemists knew nothing about the structure of the atom, electrons, neutrons and protons. But, bringing their work up to date, we can think of these periodic tables of the elements as being laid out in terms of the increasing number of protons in the nucleus, the atomic number, as it is called, that defines chemical properties. One of the most striking features of such a periodic table is the similarity between light elements whose atomic numbers differ by eight units. Hydrogen is something of a special case, with only one proton and no neutrons in its nucleus, so let's start with helium. Helium has an atomic number of 2, and its chemical properties are very similar to those of the gas neon, which has an atomic number of 10, and argon, with an atomic number of 18. There is, indeed, a whole family of elements with properties similar to those of helium; together they are called the inert gases, because they show a great reluctance to interact chemically with anything. In a similar way, carbon, atomic number 6, is the archetype for a family of elements which continues with silicon (atomic number 14), while oxygen (atomic number 8) is the lightest member of a family, whose next member is sulphur (atomic number 16).

Bohr realised that the chemical properties laid bare in the periodic table could be explained in terms of electrons if the electrons slotted in at different distances from the nucleus of an atom, in well-defined locations called shells. Hydrogen's lone electron occupies the closest position it can to the central proton, and there is just room, in a quantum mechanical sense, for a second electron to slot into the same shell to make a helium atom when the number of protons in the nucleus is increased to two. But there is *only* room for two electrons in this shell. For lithium, the element which has three protons in its nucleus, the third electron has to go into a shell which is further out from the nucleus. That shell has room for eight electrons, so as the atomic number increases new electrons are happily slotted in until we reach neon, which has atomic number 10 and a full, or closed, shell. It happens that a closed shell is a particularly stable configuration. Helium, with a closed innermost shell, is therefore very stable and does not react chemically; neon, with a closed *second* shell, is in a similarly stable configuration and also does not react chemically. But what if we add one more proton to the nucleus and one more electron to the cloud?

Sodium is the atom with atomic number 11. It has two electrons in a closed inner shell, eight in a closed second shell, and just one out on its own in the third shell. As far as the outward appearance of the atom is concerned, this is very similar to lithium, which has three protons, a closed inner shell of two electrons, and just one lone electron in the third shell. Indeed, in many ways it is rather like hydrogen, which has just one electron on display. But hydrogen is something of a special case, difficult to fit into any single chemical category. In the case of lithium and sodium the lone electron in the outermost occupied shell makes the atom very unstable, in the sense that it has a great affinity for other atoms and eagerly forms chemical compounds by combining with other atoms to make molecules. In effect, each atom 'loses' its extra electron to a partner, revealing the stable, full shell within.

The situation gets more complicated with larger atoms. There are places in the periodic table where the process seems to pause for breath; we now know that this is because some shells can contain more than eight electrons, and that after some *outer* shells have received what seems like a full complement of eight electrons, some extra electrons go into *inner* shells, producing an array of elements whose chemical properties differ very little from one another because they have identical outer shells even though they have different atomic numbers, different numbers of protons in their nuclei and different numbers of electrons in their clouds. These details are not important here, although of course they are of prime importance to chemistry and the explanation of these puzzling features of the periodic table was one of the great triumphs of quantum theory.

Like everyone else who thought much about electrons in the second decade of the twentieth century, Bohr thought of them as particles. The image of the atom he developed was of electrons orbiting around the nucleus something like the way planets orbit the Sun, but with two 'planets' allowed to occupy the orbit of Mercury, eight allowed in the orbit of Venus, and so on. The image was wrong, because electrons are not particles and they are not in orbit around the nucleus. But from this naive picture Bohr was able to develop an explanation of atomic structure in terms of quantum theory. Each orbit – what would now be called, less misleadingly, an orbital – corresponds to a definite amount of electric energy locked up in the atom, the energy corresponding to the force between negative electron and positive proton at some distance from each other. All systems tend to go into the lowest energy state that they can, and that is why it was a puzzle that the electrons didn't fall in to the nucleus, releasing energy in the process. But quantum theory explained that there was nowhere for them to go. Instead of the energy path in to the nucleus being a smooth slope down which electrons could slide, it has to be thought of as a series of steps on which the electrons can rest. The smallest amount of energy an electron can emit corresponds to one photon of light. Because light comes in packets, electrons can only move closer to the nucleus in steps, emitting one packet of light at a time. And, for quantum mechanical reasons too complex to go into here, they can never take the ultimate step into the heart of the atom, the nucleus itself.

Think of the shell structure of the atom – the orbitals – as like rungs on a ladder. Two electrons sit on the bottom rung and fill it up; eight sit on the next rung and fill it up; eight more are allowed to go on the next rung, and so on. But no electron can ever be in a state corresponding to the gap between two rungs. In the quantum world, there is no such place as 'the gap between two rungs'. And no more electrons can jump on to a rung that is already fully occupied. The rungs are called energy levels, and the spacing between them both corresponds to, and explains, the characteristic spectrum of light emitted or absorbed by atoms of a particular element. But that is another story. The problem remaining with Bohr's theory was the continuing idea of the electron as a tiny particle whizzing about the atom in some definite orbit. De Broglie's idea of electron waves, and Schrödinger's corresponding wave equation, rescued atomic theory from this misleading image and opened the way for a true understanding of the way atoms bond together to form molecules. This is the basis of chemistry.

# CHAPTER FIVE

# QUANTUM CHEMISTRY

A particle is localised in space; the only way a particle can 'surround' the nucleus is to whizz round it very rapidly. But a wave is a spread out thing. An electron wave trapped by an atomic nucleus can 'surround' the nucleus in a much more real sense, in the same way that a sound wave completely fills up an organ pipe. Such a sound wave is called a standing wave; the electron waves around an atomic nucleus can also be thought of as standing waves, and described in mathematical terms as waves trapped in the electric potential field of the nucleus. Each single electron has to be regarded as a diffuse object spread out over a volume roughly as big as the whole atom. This cloud corresponding to a *single* electron is thicker – more dense – in some places than in others. If you try to explain that in terms of particles, it 'means' that the electron 'particle' is more likely to be found in some places (where the cloud is thickest) than in others. The concept of uncertainty comes in again here. If you insist on thinking of the electron as a particle, all you can say about its position is that it exists somewhere within a particular orbital, and that it is most likely to be found in the densest parts of that 'cloud'. But it really is best to get rid of the image of an electron as a particle altogether now, and to think of the diffuse cloud around the nucleus as representing a 'real' electron.

The standing waves are described by Schrödinger's equation. This defines the shape and extent of the electron

Figure 5.1 The lowest energy state of the electron cloud around a hydrogen atom is spherical.

clouds, and they are different for different energy levels and different orbitals. But instead of thinking of the electrons in different shells as neatly outside each other, like a series of onion rings, we have to visualise them all inter-penetrating, like a lot of ripples on a pool. Every individual electron cloud extends down to 'touch' the nucleus, and all electrons come under the direct influence of the nucleus, but some more strongly than others. There are many ways of picturing what is going on. The electrons that used to be thought of as further out from the nucleus do indeed 'spend more time' further out – their orbital clouds are concentrated further from the nucleus. But the most important thing is that they are less strongly attached to the nucleus. Electrons in higher energy levels, further up the rungs of the energy ladder, are more easily detached from the atom than electrons lower down the energy ladder. And that is of key importance, as we shall see, to chemistry.

The simplest clouds are spherical and centred on the nucleus. The two electrons in the helium atom occupy these simplest orbitals, the lowest energy state. At the next level up the energy ladder, however, things are a little more complicated. The wave equation does indeed predict the existence of another spherically symmetric state, into which two more electrons with slightly more energy than the two innermost electrons can slot. But alongside it, at very nearly the same energy, there are three more standing wave patterns, shaped rather like short, fat dumb-bells, or hour glasses, at right angles to each other. Two electrons are able to slot into each of these orbitals, giving a total number of eight $(2 + 6)$ for the filled second shell. Things get still more complicated at higher energy levels. But the quantum mechanical wave equation exactly predicts how many electrons can fit into each shell, and this explains the

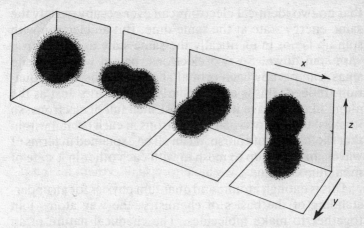

Figure 5.2 At the next level up the energy ladder, things are more complicated, and electrons can fit in to a spherical shell or any of three dumb-bell shaped shells around the nucleus. Two electrons can fit in to each of these four shells, so eight electrons fill this energy level; this explains the repeating pattern of properties of chemical elements eight numbers apart in the periodic table.

structure of the periodic table of the elements. The quantum math also tells us that although some electrons do fit spherically around their nuclei, very many electron orbitals have a definite shape and a definite orientation with respect to one another. In very many cases, electron orbitals stick out from atoms in certain clearly defined, and predictable, directions.

There is one oddity here, which calls for a little explanation and which brings in the dual particle-wave nature of the electron once again. Just when you thought it was safe to think of electrons as waves, there's a catch. Why should there be room for two electrons in each orbital? The explanation for this has to do with a property of the electron that has, unfortunately, been called 'spin', even though it bears little resemblance to the spin of an object in the everyday world, like a child's top or the Earth in space. Electrons can be slotted into orbitals in one of two spin states, 'up' or 'down'. Quantum mechanics – quantum math – predicts

that no two identical electrons can ever occupy exactly the
same energy state at the same time. But an electron with
spin up is not in identically the same state as an electron
with spin down. So two electrons, paired with opposite
spins, can occupy each orbital allowed by the wave equation. Indeed, this is a particularly stable state. Just as the
atom 'likes' to have its outermost shell full of electrons, so
it 'likes' to have two paired electrons in each orbital within
that shell. And, of course, it can all be explained in terms of
waves, matched up to mesh in with each other in a state of
minimum energy.

This is enough atomic and quantum physics for an understanding of the basics of chemistry, the way atoms join
together to make molecules. The chemical nature of an
atom depends on the number of electrons in the highest
energy shell that is occupied at all; those electrons are best
thought of as spread out, three-dimensional objects with a
definite shape, attached to the nucleus and sticking out into
space, each one covering a volume comparable to the size
of the atom itself; full shells are particularly stable, so that
atoms 'like' to arrange themselves to get filled outer shells;
and electrons come in two flavours, up and down, which
'like' to pair up with one another. It was from this basis that
Pauling jumped off to invent modern chemistry and
opened the way to molecular biology.

# Chemical bonds

Linus Pauling, the great American chemist, cut his scientific teeth in the midst of all the excitement about quantum
theory. Born in 1901, he gained his first degree, in chemical
engineering, from Oregon State Agricultural College (now
Oregon State University) in 1922, and then moved to the
California Institute of Technology, where he completed his
PhD in physical chemistry in 1925, the year that de
Broglie's concept of the electron wave began to gain cur-
rency in scientific circles. For two years, immediately following his PhD work, Pauling visited Europe on a
Guggenheim Fellowship. He worked for a few months in

Munich, in the same department as Arnold Sommerfeld; in Copenhagen at the Institute headed by Niels Bohr; with Erwin Schrödinger in Zürich; and visited William Bragg's laboratory in London. He was in the right place at the right time, the right age and with the right training, to assimilate the dramatic new developments and use them to explain how atoms join together to form molecules. He was also smart enough to put the pieces of the puzzle together, and developed into a brilliant communicator. His books on chemistry are still unequalled as introductions to the subject, both for the serious student and the layman with a passing interest.

By the middle of the 1920s, it was already clear that molecules are held together by the sharing of electrons between atoms. If each atom has an electronic grip on the same electrons, then effectively the atoms have a grip on each other. And it was also already clear that the nature of this sharing – this bonding – between atoms was related to the desirability, in atomic terms, of completed, or closed shells of electrons. The idea is best introduced by looking back to the early nineteenth century, before chemists knew that electrons existed, when it was first realised that there are two types of chemical compound. One type, such as the compound between sodium and chlorine which we call common salt, can be dissolved in water to produce two types of electrically charged ion. The symbol for sodium is Na, that of chlorine is Cl; when sodium chloride (NaCl) is dissolved in water and a positively charged electrode is placed in one end of the solution and a negative plate at the other end, negatively charged chlorine ions ($Cl^-$) move towards the positive plate and bubble off as chlorine gas. (What happens at the negative plate is a little more complicated; hydrogen, from the water in the solution, generally takes up the available electrons and is bubbled off, 'stealing' the electrons that 'ought' to go to the sodium ions.) The bond between Na and Cl in molecules of NaCl clearly involves electricity, and is called an electrovalent, or ionic, bond. The pioneering nineteenth-century work on electrovalency came from the British scientist Henry Cavendish, and from the Swede Svante Arrhenius. There are, however, many compounds which, even though they may dissolve in water, do not conduct electricity and cannot be separated into positive and negative ions. Their molecules are held together by a different kind of bond, called a covalent bond.

Even before anyone had any inkling of what made a chemical bond, it was clear, as I have mentioned, that atoms of each element had the ability to make a specific number of bonds, whether they be covalent or ionic, with other atoms, Carbon, for example, has a valence of four, meaning that it can make four bonds. Hydrogen has a valence of one. So when carbon and hydrogen combine they make molecules of methane, $CH_4$, which can be represented in terms of bonds;

$$H - \overset{\displaystyle H}{\underset{\displaystyle H}{\overset{\displaystyle |}{\underset{\displaystyle |}{C}}}} - H$$

On this picture, carbon dioxide, $CO_2$, must involve *double* bonds between oxygen, with valence two, and carbon, with valence four:

$$O = C = O$$

and although there are a few difficulties with this system when it comes to explaining substances such as carbon monoxide, $CO$, or ozone, $O_3$, the enormous overall success of the explanation of the observed chemical behaviour of elements in terms of a combination of bonds and valence kept these little difficulties largely swept under the rug. By the end of the nineteenth century, the concept had been refined still further. It was clear that bonds were real entities which had a definite arrangement in space. In the case of carbon, for example, the four bonds extend out from the central atom toward the four corners of a regular tetrahedron. The discovery of the electron offered the first prospect of explaining any of this in detail.

Although the bond between sodium and chlorine in a salt molecule can be written

$$Na - Cl$$

in view of the ionic nature of the bond it is better written

$$Na^+Cl^-$$

One neutral atom has become a positive ion; the other neutral atom has become a negative ion. The two are then held together (more accurately, many such ions are held together in a regular array, a crystal lattice*) by electrostatic forces. Knowing that atoms are partly made of electrons, that electrons carry negative charge and, as experiments show, that the charge carried by each ion is the same magnitude as the charge on the electron, it is clear what has happened. One electron has left the sodium atom and become attached to the chlorine atom. But why?

Sodium has an atomic number of 11. It has 11 protons in its nucleus, and 11 electrons in a cloud around it. Two of those electrons fill the innermost shell, eight others fill the second shell, and just one lonely electron is left in the third shell. In many ways, the chemical properties of sodium resemble those of hydrogen, which has only one electron. But if, somehow, a sodium atom could get rid of its outermost electron, then what would be left behind would be two closed shells, a positively charged ion with more than a

---

* The discovery of the structure of solid sodium chloride came in 1913, with the aid of the X-ray diffraction method that was to prove a key tool in the subsequent development of molecular biology. Once you know that X-rays are waves, or can be treated as waves on appropriate occasions, you can use the exact nature of the diffraction pattern found by scattering X-rays off the atoms and molecules in a solid to work out the structure of the solid. In this case, the experiments showed that there is no such thing as a single NaCl molecule. Instead, each sodium atom is equidistant from six neighbouring chlorine atoms, and each chlorine atom is similarly equidistant from six neighbouring sodium atoms. In a sense, the whole crystal – each grain of salt – is one giant molecule.

superficial resemblance to an atom of neon, one of those very stable, non-reactive inert gases. Now look at chlorine. Chlorine has an atomic number of 17. Each atom has 17 protons in its nucleus (plus a comparable number of neutrons) and 17 electrons in its cloud. How are they arranged? Working outwards in terms of shells, we have a closed shell of two electrons, another closed shell of eight, and a very nearly full shell of seven. Six of the electrons in that outer shell are paired, up with down, to fill three orbitals. The other occupies an orbital by itself, with no partner. If only the chlorine atom could somehow steal an extra electron from somewhere, it could fill up the outer occupied shell, mimicking the appearance of the inert gas with atomic number 18, argon. That is exactly what happens. Each sodium atom donates an electron, each chlorine atom gratefully accepts an electron. Quantum mechanical calculations show that the result, in a crystal lattice of sodium chloride, is a lower energy state, and natural processes always tend to move towards the lowest possible energy state. The electrons are not so much shared by two atoms as passed completely from one atom to the other.

Covalent bonding is much more a matter of sharing, but the end result is the same – outer shells for all the atoms involved that are effectively filled, and have the most stable, inert gas configuration. Hydrogen provides the simplest example. Each hydrogen atom has one electron; ideally, it would 'like' to have two, a closed shell with the same structure as the electron cloud around a helium atom. It can achieve something very much like this – a lower energy state – by sharing its electron with another hydrogen atom, and gaining in return a half share in the other atom's electron. Both electrons belong to both atoms, and the bond can be written in a form which makes this clear

$$H : H$$

The pair of dots represents the two electrons shared between the two atoms.

In the same way, the idea of shared electrons begins to

Figure 5.3 When two hydrogen atoms combine to form a hydrogen molecule, one elongated electron cloud surrounds both nuclei.

provide insight into the nature of the bonds between carbon and hydrogen in methane:

$$H : \overset{\displaystyle ..}{\underset{\displaystyle ..}{C}} : H$$

These ideas came from the American Gilbert Lewis, of the University of California, who discussed the idea of forming ions by completing closed shells of electrons, and making covalent bonds by sharing atoms between electrons, as early as 1916. But no real further progress could be made until the development of quantum mechanics and the concept of the electron wave which provided for a full mathematical description of the behaviour of electrons in atoms and the nature of orbitals. In 1916, even the concept of electron spin had still to rear its head, and there was no clue as to why paired electrons, as well as filled shells, are important in chemical bonding.

There is not, in fact, any clearcut dividing line between compounds that are purely ionic in nature and those that are purely covalent. Some are more ionic in nature, others seem predominantly covalent, with all shades of grey in between. All bonds can be regarded as a mixture, a blend of the two types. Even hydrogen can be thought of as making ionic bonds. If one atom gave up its electron to the other, we would have a molecule $H^+H^-$; if the exchange went the other way round, we would have $H^-H^+$. The covalent sharing of the two electrons can be thought of as very rapid switching between these two states, with first one atom and then the other laying claim to *both* electrons, and the two atoms being held together by the electrostatic forces. This idea of oscillating between different states, or resonance, was at the heart of Pauling's work in the late 1920s, the work which put the theory of covalent bonding on a secure mathematical footing within the framework of quantum theory. As far as molecular biology is concerned, the covalent bond is *the* bond. Purely ionic bonds like those in sodium chloride play no part in forming the molecules that are important to life, although resonance is important and weaker electrostatic forces are very important in determining the exact shape of large biomolecules. By explaining the nature of the covalent chemical bond, Pauling began to explain the chemistry of life.

# The covalent bond

The rules based on paired electrons and developed by Gilbert Lewis were empirical – rules of thumb that gave an idea of what was going on but lacked a secure basis in physical theory. Within a year of Schrödinger publishing his quantum mechanical wave equation, two German physicists, Walter Heitler and Fritz London, had used this mathematical approach to calculate the change in overall energy when two hydrogen atoms, each with its own single electron, combine to form one molecule with a pair of shared electrons. The change in energy depends on the re-arrangement of the electrons within the electric field produced by the two nuclei.

This is like the change in the energy of an object – such as a ball – when it moves about in the gravitational field of the everyday world. A ball resting on a table has more gravitational potential energy than the same ball resting on the floor. Push it to the edge of the table, and it will fall off. Nature always seeks the lowest energy state. Thanks to the equations of quantum mechanics – not just Schrödinger's equation, although that was the one which most physicists found most convenient for this kind of calculation – it is possible to calculate how much energy the electron has in each orbital state, in an analogous way to the method by which physicists calculate the difference in energy between the ball on the table and the ball on the ground in the everyday world. It is, indeed, *differences* in energy that are all important. And when Heitler and London calculated the difference in energy between two hydrogen atoms and one hydrogen molecule, they came up with a number very close to the amount of energy which chemists already knew, from experiment, was required to break the bonding between the atoms in a hydrogen molecule. Later calculations, including refinements made by Pauling, have given even better agreement. Just as a physicist of the old school can calculate how much energy it takes to lift the ball from the ground and on to the table, so quantum physics tells us, exactly, how much energy it takes to split a hydrogen molecule into its constituent atoms. This was a really startling development in 1927. Instead of just saying that, for reasons unknown, electrons like to pair up and atoms like to have filled electron shells, physicists were now able to calculate the change in energy when electrons paired up and shells were filled. The calculations confirmed that there was no arbitrariness about the arrangement, and that the arrangements which are most stable in the atomic world are always the arrangements with the least energy. In Pauling's own words, 'the covalent bond consists of a pair of electrons shared between two atoms, and occupying two stable orbitals, one of each atom'.*

## Carbon's hybrid bonds

In principle, the same calculation can be applied to any molecule; in practice, this becomes very difficult for complex molecules, and various approximation techniques have to be used. I shall gloss over these difficulties,

---

* *Chemistry*, page 143.

however, and concentrate on the key concepts with which Linus Pauling opened the door to a proper, quantitative study of the nature of biological molecules. The most important atom in biology is carbon, and the study of carbon chemistry is so important that it is treated as a separate branch of chemistry, organic chemistry. Superficially, the reason for this is simple. Carbon has a valence of four – it can form four bonds with other atoms. Because the most stable configuration for an atom is a filled outer shell of eight electrons, this is normally the largest valence any atom can have. With less than four electrons in its outer shell, an atom will tend to form molecules in which the electrons (up to three of them) are in effect given up to its partners, laying bare the filled shell beneath; with five, six or seven electrons in an outer shell, an atom has room for only three, two or one partners. Carbon has four electrons, each unpaired in an orbital, ready and willing to latch on to electrons from other atoms. There are, of course, other atoms in a similar configuration. Carbon has atomic number six, two electrons in its innermost shell and four in the chemically active shell; silicon has atomic number 14, with its electrons in a 2:8:4 configuration. Like carbon, it forms four bonds at a time. But, because the four key electrons are in a higher energy state ('farther out from the nucleus' on the old picture) the bonds are weaker.*

In the late 1920s, the new success of the orbital theory and quantum mechanics in explaining the energy of the hydrogen molecule clearly pointed the way to a new understanding of the covalent bond. But the path towards an

---

* As ever with nice, simple rules there are exceptions. Phosphorus, for example, has atomic number 15 and five electrons in its outermost occupied shell. It is capable of utilising each of the five electrons to make a total of five bonds at a time; the easiest way to think of this is as four covalent bonds involving shared electron pairs, as in carbon, plus one ionic bond, involving the transfer of an electron from the phosphorus to its partner. Since each of the five bonds is the same as each of the others, however, this image requires imagining each of the bonds to be 80 per cent covalent and 20 per cent ionic in character. This is very much in accordance with the new understanding of quantum chemistry developed in the 1930s.

understanding of carbon compounds, the molecules of life, was blocked at the outset by one very curious puzzle. As I have already explained, the second electronic shell, in carbon as in other atoms, consists of one spherical orbital, similar to the innermost orbital, and three other orbitals at right angles to each other. Chemists knew, from studies of the shapes of crystals formed by different compounds, that the four carbon bonds also had a three-dimensional structure, a definite alignment in space relative to the central nucleus. Logically enough, the shape and structure of a crystal reflects the shape of the molecules which make up the crystal, and although I cannot go into details there is nothing conceptually very difficult about this kind of work. But whereas the four electron waves described by quantum physics 'ought' to consist of one with no particular alignment and three perpendicular to one another, the real carbon bonds are unambiguously shown by these studies to be completely symmetrical, four identical bonds aligned towards the corners of a regular tetrahedron. Pauling explained why, in a key paper on the nature of the chemical bond, which appeared in 1931.

The explanation depends entirely on a quantum physics view of the electrons and their behaviour. Casting aside any idea of little hard particles orbiting around the atom, and building from the concept of an electron as some hybrid of particle and wave, Pauling hit on the idea of each of the four symmetric orbitals in the atom as a hybrid combination of the four fundamental orbital states. These orbitals come in two different varieties, which were identified and labelled $s$ and $p$ on the basis of spectroscopic studies long before anyone had any idea that such things as spread out electrons occupying volumes of space with distinct shapes existed. Serendipitously, it happens that these initials provide a convenient mnemonic for remembering the shapes of the orbitals. The spherical, or $s$, orbital had to be mixed in with the three perpendicular, or $p$, orbitals, to produce four orbitals denoted as $sp^3$. Just as it is impossible to say whether an electron is 'really' a wave or a particle, so it is impossible to say whether a particular bond is 'really' $s$

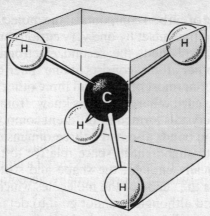

Figure 5.4 Electrons in the spherical, *s*, orbital ought to behave differently from those in the perpendicular, *p*, orbitals. But an atom like carbon forms four identical bonds. Each of these is a hybrid, flavoured with one part of *s* and three parts of *p*. Such behaviour is only possible because electrons behave as waves, not particles, in these interactions.

or *p*. It is both, at the same time, in a ratio 1:3.* Although the idea jumped off, in empirical fashion, from the crystal studies which showed the symmetry of their carbon bonds, it was born out by the quantum math. The symmetric state is, indeed, one with lower overall energy than a state of one pure *s* and three pure *p* orbitals. If you want a physical picture of why this should be so, it is because the four hybrid orbitals keep the four electrons, or electron clouds, at the greatest separation from one another. As you know, like charges repel; the electrons (harking back to the picture of little negatively charged particles) would 'like' to be as far apart from each other as possible, and hybridisation of the available orbital states allows them to achieve this.

But Pauling didn't stop there. The same concept of hybridisation of quantum mechanical states, resonance between one form and another like the resonance which

---

* The same concept is required to understand the ability of a phosphorus atom to form five identical bonds simultaneously.

makes the electron both a particle and a wave at the same time, led him to an explanation of other fundamental features of organic chemistry.

# Resonance

This relates to the image of the $H_2$ molecule as a resonance between $H^+H^-$ and $H^-H^+$, which Pauling had developed in 1928. The principle of resonance says that if a molecule can be described in two (or more) equally acceptable ways (where 'acceptable' effectively means states with the same energy, different versions of the lowest possible energy state for that molecule) then the molecule has to be thought of as existing in both (or all) of those states simultaneously. The 'real' molecule is a hybrid of all the possible structures with the same lowest energy, just as the 'real' carbon orbitals are mixtures of $s$ and $p$ states. The example of ozone, a relatively simple molecule, gives an idea of what is going on, and is especially interesting because it brings out other features of chemical bonding at work.

Oxygen has a valence of two, with six electrons in its outermost occupied shell. It is easy to write down (ignoring the inner, filled electron shells) representations of molecules such as water, $H_2O$, and oxygen, $O_2$, on the old picture of shared electrons:

$$H - O - H$$

$$O = O$$

Each bond, indicated by a dash, represents a pair of shared electrons. But how do you account for the fact that oxygen also forms a tri-atomic molecular form, ozone, $O_3$? The first step is to appreciate that covalent bonding is not the whole story, and that the ionic transfer of an electron from one atom to another ought also to come under consideration, as in the hydrogen molecule. One way this could happen is for one oxygen atom to give up an electron to another oxygen atom. The first is left with a net positive charge and an outer shell containing five electrons, very similar to the structure of an atom of nitrogen. It now has the capacity to make three covalent bonds. The other atom has gained an electron and a net negative charge; it has an electronic structure like that of fluorine and room for only one more electron in a covalent bond, an outer shell of seven electrons. If we follow the convention of using a dash to represent a pair of electrons in a bond, and dots to represent the electrons that are not part of the covalent bond, this gives us the two following ways to represent the structure of an ozone molecule:

Spectroscopic techniques, which measure the amount of energy radiated or absorbed by a molecule, provide a direct measurement of how much energy is stored in each bond. A double bond is stronger than a single bond; it provides a tighter link between atoms and holds them closer together. So, if this picture is correct, there ought to be evidence of two different bond energies, corresponding to two different bond lengths, apparent in the spectrum of ozone. There isn't. Instead, the spectroscopic studies show quite clearly that ozone molecules are held together by two equal bonds, each equivalent to a bond strength of 1.5. The explanation is that the 'real' structure is a resonance between the two possibilities outlined in the diagrams, a hybrid structure like the hybridisation which gives us the $sp^3$ orbitals in the carbon atom. Unlike that kind of hybridisation, however, in this case the picture does involve a re-arrangement of electronic charge which produces asymmetry. One atom has effectively lost an electron, and has a resulting surplus of positive charge in its vicinity; the other two atoms have each effectively gained half an electron, with a corresponding increase in the negative charge at either end of the molecule. Weak net charges of this kind commonly arise in molecules, especially large molecules which contain many atoms. Because opposite charges attract one another, while like charges repel, this leads to a tendency for large molecules to stick together in certain ways, and, indeed, for different bits of very large molecules to stick together in a weak form of electrostatic bonding. As we shall see in Chapter Six, this turns out to be of great importance in the molecules of life.

The possibilities are almost endless. One very common substance whose structure depends on resonance and hybridisation is the carbonate ion, $CO_3^{--}$, found in common chalk, the shells of sea creatures, and limestone. Even though it consists of four atoms, the carbonate ion is a stable structure which operates as an entity in many chemical processes. It has a double negative charge because it has taken up two electrons from other atoms which are eager to get rid of them and form ionic bonds, and clearly since the $CO_3^{--}$ structure is so common in nature it must be very stable in energy terms, in terms of closed electron shells. How can you arrange four electrons from the carbon atom, six from each of the oxygen atoms, and the two extra electrons in the most stable state?

There are three possibilities, variations on the same theme like the two ozone variations on the tri-atomic oxygen theme. Two of the three oxygen atoms have gained an electron and are left with one 'hole' suitable for making a covalent bond with carbon. The other oxygen atom can form the usual double bond.

The three possibilities are equivalent and have the same energy as each other. Because the bonds are different, each of those three possibilities would be asymmetrical, and the asymmetry would show up in spectroscopic measurements. Once again, however, the measurements show that the carbonate ion is perfectly symmetrical. Three bonds, each equivalent to 1.333 normal bonds, are arranged

at 120° to one another uniformly around the carbon atom.
The carbonate ion is a resonance hybrid.

It is crucial, however, that the different states involved in
the resonance have the same energy. If a molecule can exist
in two or more configurations which have different energy,
then there is no resonance and it does, indeed, form two
*different* compounds. A simple example is dichloro-
ethylene, a compound in which each molecule contains two
carbon atoms, two hydrogen atoms and two chlorine
atoms. The chemical formula may be $C_2H_2Cl_2$ in both
cases, but there are two structural formulas.

The left-hand version is called the *cis* isomer; the right-
hand version is the *trans* isomer. They have different densi-
ties, different melting and boiling points, and other different properties. The differences are produced by the
differences between hydrogen atoms and the much larger
chlorine atoms; in the *cis* version, the large chlorine atoms
on the same side of the molecule interfere with each other;
in the *trans* version, the molecules are neater structures
with the large chlorine atoms kept apart. To change one of
them into the other would require spinning one end of the
molecule over, and you can't rotate a double bond.

In very many organic compounds the basic unit is not the
carbon atom itself but a group of six carbon atoms holding
hands in a ring. This is called the benzene ring, because the
simplest such molecule consists of six carbon atoms and six
hydrogen atoms, a molecule of benzene, $C_6H_6$. The structure can be written two ways, each with double bonds
between alternate pairs of carbon atoms:

By now, it should come as no surprise to learn that all chemical studies of the strength of the bonds in the benzene ring show that each 'real' bond has a strength of 1.5. The actual benzene ring is another resonance hybrid. This very stable structure forms the basis of an enormous number of molecules, including very many of the molecules of life. Slightly more complex variations on the theme can be made by replacing one or more of the hydrogen atoms around the ring by groups of atoms such as the methyl group, $-CH_3$. Groups like this, denoted by $-R$, can replace hydrogen atoms wherever there is room to do so, which generally means on the outside of a large molecule. Drawing the ring itself simply as a hexagon, leaving the carbon atoms and single hydrogen atoms as understood, we get compounds such as toluene:

and, a little more complicated still, trinitrotoluene ('toluene with three nitro groups added'), or TNT:

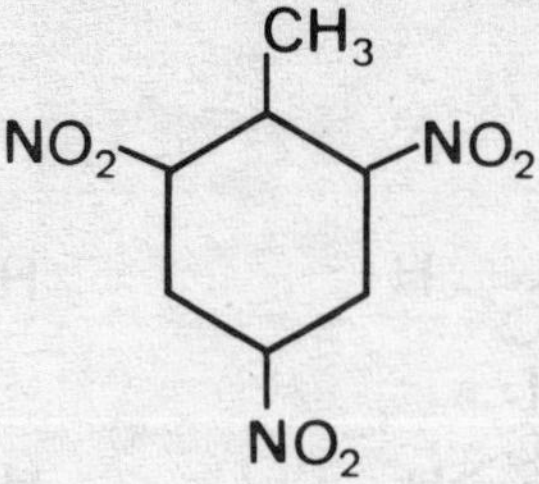

A generalised shorthand to denote families of molecules with the same sort of structure as toluene would be

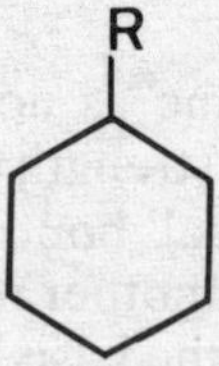

where different groups (R) can attach to the basic benzene ring. In all of these molecules, resonance hybridisation is an important factor in explaining the structure. And, as well as the family of compounds built around the single benzene ring, there are many more families, of increasing complexity, based on two or more rings joined together. They, and the compounds based on a single benzene ring structure, are called the aromatic hydrocarbons; the simplest two-ring version is naphthalene, $C_{10}H_8$:

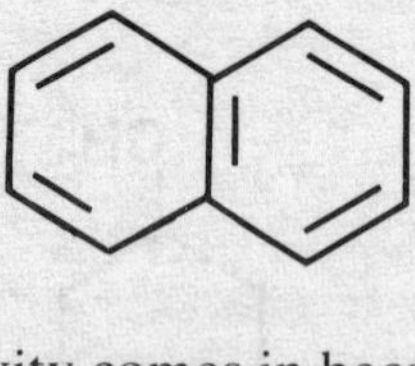

Now, a further complexity comes in because the rings can join together in different ways. Each ring is held flat by the double (or, if you like, 1.5) bonds, but there may be more than one way to join another ring on to an existing chain. Anthracene and phenanthrene, for example, are both written in shorthand $C_{14}H_{10}$, but have different structures:

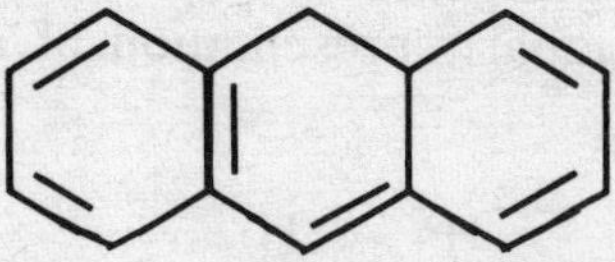

Anthracine 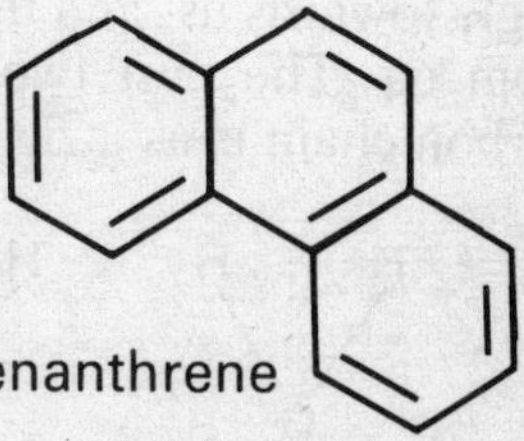                    Phenanthrene

The basic benzene structure in all of these molecules is the resonance hybrid, which, the quantum math confirms, is energetically more stable than either of the two variations on the theme represented by alternating single and double bonds around the ring. This, however, is far from being the end of the carbon chemistry story.

# Polymerisation

Carbon atoms are quite happy forming covalent bonds with other carbon atoms – that, indeed, is the basis of the benzene ring structure. The benzene ring is a particularly stable molecular form, in terms of energy, because the natural angles made by the four bonding carbon orbitals comfortably fit such a six-sided structure. But a string of carbon atoms can also form the backbone of a long chain of atoms, a polyatomic molecule. The simplest versions are variations on the methane theme:

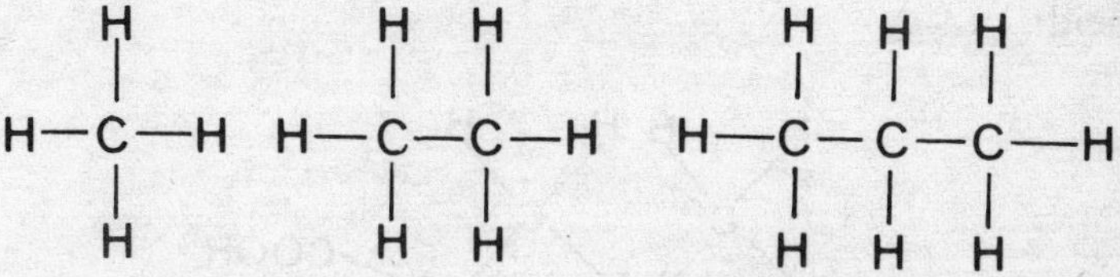

and so on. It is often easiest to draw these long molecules as straight chains, but remember that in reality the carbon bonds point towards the corners of a tetrahedron. If we draw two of these bonds in the plane of the paper, we have to imagine the other two sticking either up or down from the carbon atom, with one coming out of the paper at an angle towards us, and the other going into the paper away from us. The best two-dimensional representation of a carbon chain is as a zig-zag:

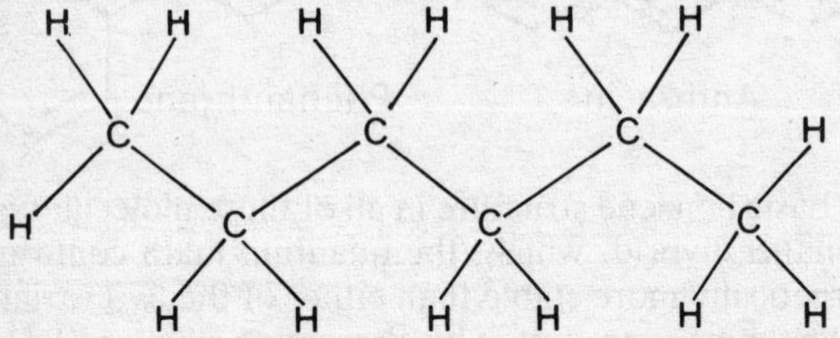

A slightly more complex structure might have a hydrocarbon chain with one multi-atom group at each end, like diaminohexane ('di' meaning 'two', 'amino' being the name of the $NH_2$ group, and 'hexane' because it contains six carbon atoms. In other words, a six-atom carbon chain with two amino groups joined to it.)

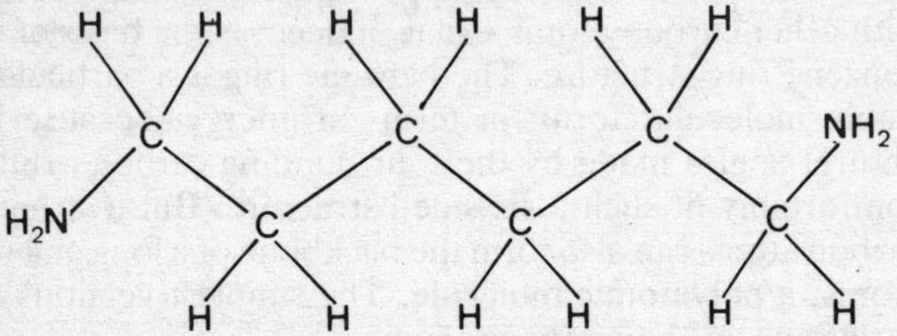

Another molecule with a very similar structure is adipic acid:

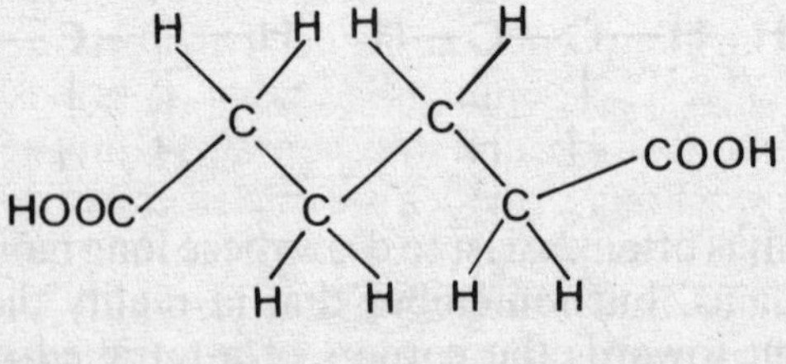

and when you combine these two molecules together, under the right conditions, a $-COOH$ group from the adipic acid can release $-OH$ while the $-NH_2$ group on diaminohexane releases $-H$. The released atoms combine to form water, $H_2O$, while the ends of the two carbon-based molecules join up to make a longer chain:

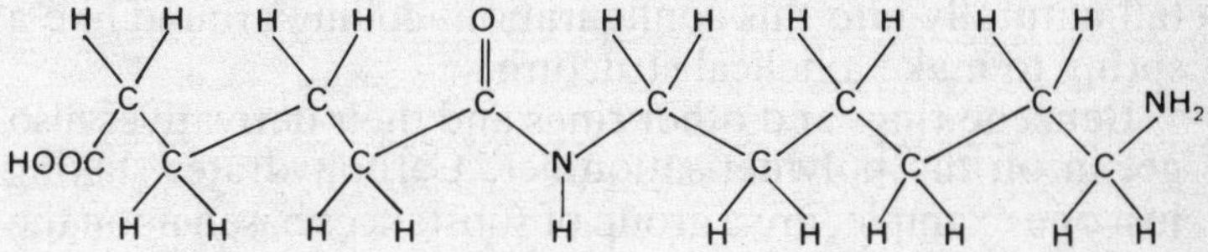

Of course, the process need not, and does not, stop there. Other adipic acid molecules and diaminohexane molecules can join on to the chain at the appropriate ends, each time releasing a water molecule. The process by which complex molecules join together and eliminate a simple molecule, such as water, is called condensation; the resulting long chain molecule, which may contain thousands of basic units in a repeating sequence, is generally called a polymer. The one outlined here is particularly useful – it is called nylon. And virtually any of the hydrogen atoms sticking out from the spine of the polymer can, in principle, be replaced by a group, such as $-NH_2$, or even by a more complex structure such as another carbon-based chain, or a benzene ring type of structure.

Such chains are not rigid structures, like a stick or a pencil. The carbon bonds have plenty of play in them, allowing a considerable degree of bending. The molecules can get tangled with each other, and one long molecule of this kind can bend back upon itself to form a complex, knotted shape. This bending and tangling brings different atoms on the side groups into contact with one another and, as we shall see, provides the opportunity for the weak electrostatic forces I have already mentioned to come into play. It can also produce a regular structure, very often with stretches resembling a series of benzene rings stacked on top of each other and slightly displaced from one another. The natural angle between the carbon bonds makes the benzene ring shape particularly favoured, and in a long carbon chain the same natural angle can make the chain tend to loop round on itself. In this case, though, the carbon atoms are not joined to close the ring, but continue the polymer chain around, looping on top of the ring below, like the coils of a snake. A long run of polymer may fall naturally into this configuration, coiling around like a spring to make a helical structure.

Benzene rings and other rings and their derivatives also get in on the polymerisation act. Carbohydrates, taking just one example, are a group of substances based upon the familiar ring structure. Most of the carbon atoms are joined to two other carbon atoms, as in the benzene ring, but have each of their other two bonds used in combination with other atoms or groups, $-OH$ on one side and $-H$ on the other (together, of course, these groups would form water; the term 'carbohydrate' literally means 'watered carbon'). The simpler carbohydrates are called sugars – one-ring systems are called monosaccharides, and double-ring versions are dubbed disaccharides – while more complex ones, consisting of many rings joined together, are called polysaccharides. The way it works can be seen from an example which starts out from the simple sugar glucose, a monosaccharide. In glucose, one of the carbon atoms in the ring has been replaced by oxygen, and one of the $-OH$ groups attached to another of the carbon atoms has been replaced by the more complex group $-CH_2OH$. Its structure can be indicated in schematic form, if you imagine the plane of the ring as perpendicular to the page and the side atoms, or side groups, sticking up or down, above or below the ring:

Glucose

It ought to be fairly easy to see how two of these rings can get together. An −OH group at the end of one ring combines with the hydrogen atom in the −OH group on the end of the other ring, and is eliminated as water. That leaves one oxygen atom with its two bonds, −O−, forming a bridge between the two rings, together making one molecule of the disaccharide maltose.

Maltose

Just as in the case of the nylon polymer, however, the process need not stop there. Very many glucose units can be combined in this way to make long polymers of a polysaccharide. The particular polysaccharide formed in this way is called starch; a slightly different arrangement of the glucose units in a similar long chain forms the basis of another familiar biological substance, cellulose.

In these examples, there are six carbon atoms in each basic monosaccharide unit. Some monosaccharides, however, contain only five carbon atoms, four of which plus one oxygen atom form a five-sided ring, while the fifth is part of a side group, −CH₂OH. These compounds are called pentoses. One of them, exactly like glucose except for the missing carbon atom and its associated side groups, is called ribose. Another, similar to ribose except that one of its −OH groups has lost the oxygen atom, leaving a simple C−H bond behind, is called, logically enough, deoxy-ribose (meaning 'ribose from which one of the oxygens has gone'). It is the basic unit that provides the name for deoxyribonucleic acid, or DNA.

In moving upward from simple compounds to more complex molecules we have arrived, like the physical chemists of the 1930s, at the edge of life. Molecules only a little more complicated than the ones I have just described, made of carbon-based chains and rings joined together and with different groups attached to them, are the molecules of life, amino acids, proteins and the rest. For providing the basic understanding of all of these complexities in his work in the late 1920s and early 1930s, Pauling received the Nobel Prize for chemistry in 1954, the citation reading 'for his research into the nature of the chemical bond and its application to the elucidation of the structure of complex substances'.

Seldom can such an award have been more appropriate; Pauling also received a Nobel Prize for peace in 1963 (the 1962 Prize), with no official reason given although it is assumed this was a reward for his strenuous efforts to persuade politicians to ban atmospheric testing of nuclear weapons, a ban which actually came into force that year. Both honours were well earned and place Pauling in the highest rank, not just of scientists but of human beings. And yet, although his work on the chemical bond forms the whole basis of modern chemistry and molecular biology, and would be enough of an achievement to satisfy almost any scientist, it merely represented the first major development in a long career. The surprise is not that Pauling received one Nobel Prize, or even two, but that his later work investigating the nature of the molecules of life did not lead to similar recognition in a third category. As well as Pauling's own studies of biological molecules, his techniques and methodology provided the guiding light for later generations of molecular biologists, including the discoverers of the double helix itself. His first great achievement takes us to the brink of life; his next major contribution to science involved the molecules of life themselves.

# CHAPTER SIX

# MOLECULES OF LIFE

'By 1935', says Pauling, 'I felt that I had an essentially complete understanding of the nature of the chemical bond.'* Armed with this thorough knowledge of the way individual atoms join together to form simple molecules, it was only natural that his attention should have turned to ever more complex molecules – the molecules of life. The step that took him across the border involved a study of the magnetic properties of hemoglobin, the molecule that carries oxygen around in your blood. Hemoglobin is one example of a protein which is combined with another substance (in this case, iron). Such a molecule is called a conjugated protein. The iron atoms at the heart of the protein ball are responsible for holding on to, then releasing, the oxygen atoms required by the body. In the mid-1930s, E. Bright Wilson, a research student at CalTech, was working for his doctorate under the guidance of Pauling, investigating magnetic properties of several compounds. The nature of the magnetic properties of the compounds ought to depend on the way the electrons in the molecules are arranged in orbitals, and the experiments were so successful, producing results exactly in line with the new understanding of chemical bonds, that Pauling encouraged other researchers at CalTech to look at hemoglobin in the same way.

---

* Judson, *The Eighth Day of Creation*, page 77.

The immediate outcome of this work was the discovery that all the oxygen electrons form pairs when oxygen combines with hemoglobin to form oxyhemoglobin; in addition, the studies showed that iron has four unpaired electrons in hemoglobin, but none are left unpaired in oxyhemoglobin. Interesting stuff to chemists, if a little esoteric for most of us, this led on in turn to studies of many other life molecules using the same technique. More relevantly to the present story, however, the hemoglobin research was noticed by many other biologically-oriented scientists, including Karl Landsteiner, at the Rockefeller Institute for Medical Research. He encouraged Pauling to apply the new understanding of chemical bonds to problems involving the structure of other molecules of life. Set firmly on the biological side of the biochemical fence, and with hemoglobin the first biomolecule that he had been involved in investigating, it is hardly surprising that Pauling soon turned to the puzzle of how the polypeptide chain, of which hemoglobin is made, gets to be folded up into a ball, and, indeed, how other proteins fold up to form their characteristic molecular shapes. This work really began in 1937; to set the scene, we should recap briefly on just which atoms and molecules are important to the study of life.

## Building blocks

Even at the atomic level, the living world is different from the world as a whole. There are 92 chemical elements which occur naturally on Earth; only 27 of these are essential components of living things, and not all of these are essential to all living things. Furthermore, the proportions of these atoms of life found in living things are not the same as the proportions in which they exist over the Earth as a whole. Leaving aside water, which makes up more than three-quarters of the weight of most living things,* more

---

* The presence of so much water is, of course, a clear indication that life arose in the sea initially. We land-based life forms have to carry our own 'seas' around with us, inside our cells, in order to provide a liquid medium in which the chemical reactions of life can continue.

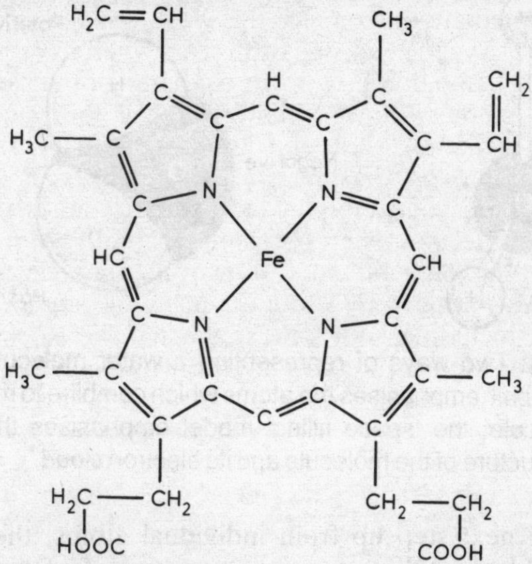

Figure 6.1 The heme group, a component of both hemoglobin and myoglobin, has an iron atom at its centre. An oxygen molecule can attach to the iron atom and be carried around the body in blood, or stored in tissue.

than half the weight of your body (its 'dry weight') is carbon, a quarter is oxygen and nearly 10 per cent nitrogen. For comparison, about 47 per cent of the Earth's crust is oxygen, 28 per cent silicon, and some 8 per cent aluminium, combined together to make rocks. Biologically important atoms have been selected by evolution because of their chemical properties – essentially, the way they form chemical bonds – which in turn give biomolecules the special properties of life. Most biomolecules are compounds of carbon; the other element strikingly over-represented among living things is nitrogen.

In terms of dry weight, proteins are by far the most important molecules of life, dominating the make-up of your body, as I have already mentioned. And proteins contain about 16 per cent of nitrogen, an even higher proportion than in the body as a whole. Many proteins are very large, complex molecules, but like all biomolecules they are built up from simple units and subunits.

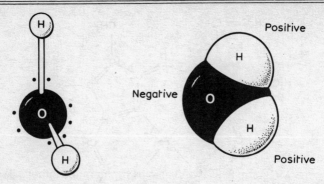

Figure 6.2 Two ways of representing a water molecule. The 'stick and ball' emphasises the atoms which combine to make up the molecule; the 'space filling' model emphasises the integrated structure of the molecule and its electron cloud.

At the next step up from individual atoms, the basic building blocks of life are units, consisting of a few atoms, which are frequently found in more complex structures. These include molecules such as ammonia, from which one hydrogen atom may be lost to form an amine group joined to a carbon chain:

and the carbon atoms themselves, as well as forming benzene rings, often crop up in many other configurations, a few of which are shown below, where the unlabelled bonds may join up with any other group R, such as the amine group, another carbon chain, or a carboxylic acid group, denoted by −COOH. The structure of the COOH group can be represented either like this:

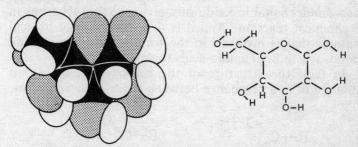

Figure 6.3 Biomolecules can also be represented either in terms of the atoms of which they are made or in terms of the distribution of charge in their electron clouds. The lumpy space filling model gives a better 'feel' for the nature of, in this case, glucose.

emphasising the importance of the single hydrogen atom at the end of the chain, or like this:

indicating that the OH group acts as a unit in some chemical reactions.

The simplest definition of an acid is a substance that readily gives up a hydrogen ion which combines with an OH (hydroxyl) group from another substance (base) to give water, $H_2O$. The carboxyl group acts in this way in many chemical reactions, which is why it is also called the carboxylic acid group, and it is this group which provides the acidity in amino acids, and in other organic acids. One of the simpler members of the carboxylic acid family is acetic acid, which is the principal ingredient in vinegar.

Confusingly, however, the carboxyl group can also act as a

base, under suitable conditions, giving up the OH group in a chemical reaction. And it is this reaction, not its acid nature, that is involved in the formation of the peptide bonds which hold amino acids together in chains.

In fact, the structure of the carboxyl group is best thought of as a resonance between two alternative forms.

The version on the left contributes 80 per cent of the character to the bonding structure, the version on the right 20 per cent. The quantum properties of the electrons which make up the bonds are crucially important in determining the overall structure of more complex molecules which contain this kind of group, and in determining details of the way the group reacts chemically. But when writing out approximate structural formulas for more complex molecules which include the carboxyl group it is usually easier to use the shorthand COOH, with the detailed structure of that particular group taken as read. In the same way, chemists commonly use the shorthand forms of other common groups, such as $-NH_2$, $-CH_3$ and so on.

Proteins are made of amino acids, and the amino acid building blocks of life all have the same basic, simple structure built around the four bonds of a carbon atom. One bond attaches the carbon atom to a hydrogen atom. Another joins it to an amine group ($NH_2$), and a third to a carboxylic acid group (COOH), giving the structure the name amino acid. That leaves one bond free to attach to another carbon based group (or, in the simplest amino acid, glycine, to a hydrogen atom) which we can denote by the letter R in a general representation of the amino acid structure:

# Amino acids and proteins

Chemists know that proteins are made of amino acids, because when proteins are boiled in a strong solution of an acid or a base the chemical bonds holding the amino acids to one another are broken, and they are left with a stew of amino acids. The names given to the common amino acids reflect the way in which they were first identified – asparagine, the first to be identified, in 1806, was initially isolated from asparagus; glycine, obtained first in the 1820s from the protein gelatin, gets its name because of its sweet taste; and so on. Very many amino acids can exist in principle – and many have been made in the lab – but, as I have mentioned, only 20 are found in all proteins. The last of these, threonine, was not identified until 1938, and there was, therefore, no hope of discovering the exact, detailed structure of protein molecules until the 1940s.

The 20 amino acids represented in Figure 6.4 are the building blocks of life. They are present in all proteins. In addition, there are two amino acids that are found in a very few proteins, and one of the common amino acids, cysteine, very readily combines with another molecule of cysteine to make cystine:

Two molecules of cysteine

One molecule of cystine, plus a released hydrogen molecule

More concisely, since the group

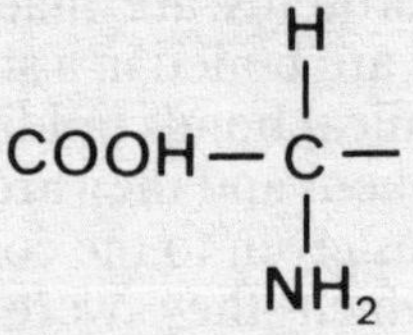

$$COOH - \overset{\displaystyle H}{\underset{\displaystyle NH_2}{\overset{\displaystyle |}{\underset{\displaystyle |}{C}}}} -$$

is the same in all amino acids, we can ignore it and concentrate on the side chains which distinguish one amino acid from another:

$$- CH_2 - HS + HS - CH_2 -$$

combine to make

$$-CH_2 - S - S - CH_2 - \; + \; H_2$$

The link between the two half-molecules is a disulfide bridge, and is very important to an understanding of the structure of proteins. It is, however, to some extent a matter of choice whether you regard cystine as a different building block of life, or simply another version of cysteine. So you may find different sources referring to 20, 21 or 23 vital amino acids, or simply, more cagily, to 'about 20'. I shall stick with the 20 shown in the figure.

The obvious question, if amino acids are the essential building blocks of the molecules of life, is: where did the first amino acids come from? During the 1920s, the English biologist J. B. S. Haldane and the Soviet researcher A. I. Oparin independently came up with the same idea. They suggested that when the Earth was young the energy supplied by lightning discharges in its violently active

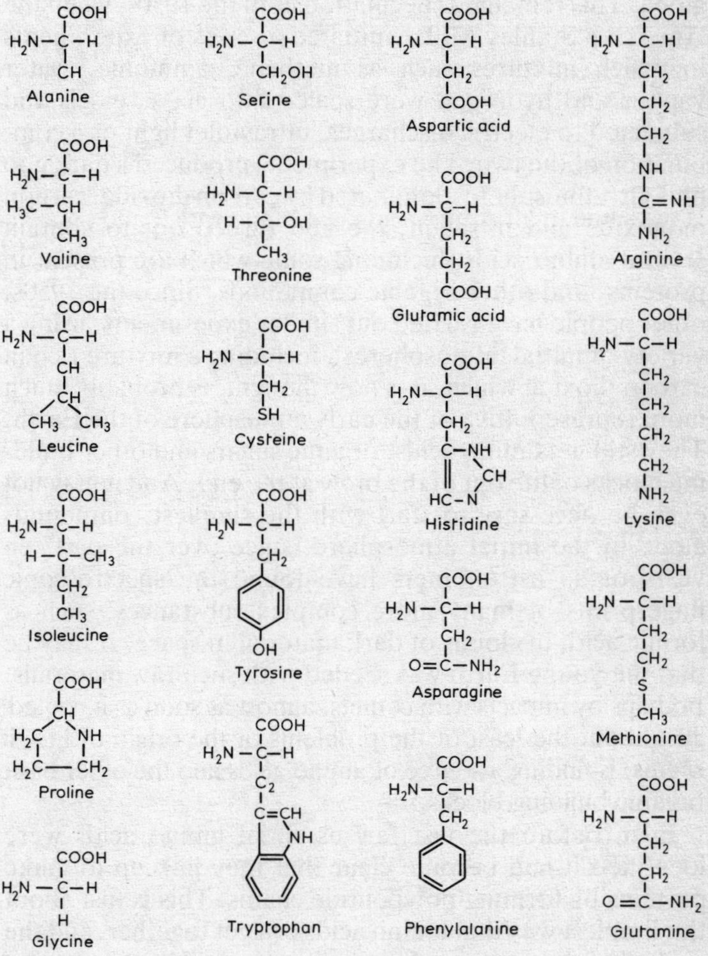

Figure 6.4 The 20 amino acids that are the building blocks of life.

atmosphere, or the heat from volcanoes, could have been
sufficient to energise chemical reactions among simple substances leading to the production of amino acids in the
oceans, or in pools of water on the cooling surface of the
globe. This remained speculation until the 1950s, when the
American Stanley Miller initiated a series of experiments
in which mixtures such as methane, ammonia, water
vapour and hydrogen were sealed into glass vessels and
subjected to electric discharges, ultraviolet light or a combination of the two. The experiments produced a dark goo
plus an 'atmosphere' dominated by carbon dioxide, carbon
monoxide and nitrogen; the goo turned out to contain
several amino acids, including some which are present in
proteins, and other organic compounds. Since the 1950s,
other people have carried out similar experiments, using a
variety of initial 'atmospheres', including a mixture rich in
carbon dioxide which, it is now thought, is probably much
more representative of the early atmosphere of the Earth.
They still get amino acids, organic sugars and other building blocks of life out of the brew at the end. And it may not
even be necessary to start with the simplest compounds
alone in the initial atmosphere, since over the past ten
years or so astronomers have found the spectroscopic
fingerprints of many more complex substances, such as
formic acid, in clouds of dark material in space. It may be
that the young Earth was seeded with such raw materials,
perhaps by impacts with comets, almost as soon as it cooled.
Just about the least of the problems of the origin of life, it
seems, is finding a source of amino acids and the other basic
organic building blocks.

Even before the last few essential amino acids were
identified it had become clear that they link up to make
proteins by forming polypeptide chains. This is just about
the simplest way that amino acids can get together, and the
groundwork for this understanding was laid by the work of
the German chemist Emil Fischer in the first decade of the
twentieth century, although in fact the proof that proteins
are polypeptide chains, and not more complex structures,
came only in the 1930s. The chains – like polymers – are
formed by condensation, eliminating a simple molecule as

two amino acids join together. It is exactly the way in which
diaminohexane and adipic acid join together (which, of
course, is why I used that example in the previous chapter).
The amine group attached to one amino acid gives up a
hydrogen atom; the carboxylic acid group on the other
amino acid gives up an OH group. A new bond – the
peptide bond – forms between the −CO on the end of one
amino acid and the −N on the end of the other, to make a
dipeptide, a union of two amino acids:

$$R-\underset{\underset{H}{|}}{\overset{\overset{NH_2}{|}}{C}}-\overset{\overset{O}{\|}}{C}-\boxed{OH} \qquad \boxed{H}-\overset{\overset{H}{|}}{N}-\underset{\underset{R}{|}}{\overset{\overset{O=C-OH}{|}}{C}}-H$$

combine to produce

$$R-\underset{\underset{H}{|}}{\overset{\overset{NH_2}{|}}{C}}-\overset{\overset{O}{\|}}{C}-\overset{\overset{H}{|}}{N}-\underset{\underset{R}{|}}{\overset{\overset{O=C-OH}{|}}{C}}-H \quad + \quad H_2O$$

Notice that the two groups R need not be the same, and
that the new molecule is bent. In fact, of course, all the
carbon bonds are angled in the way Pauling explained in
terms of hybrid orbitals. We just draw some of the chains in
straight lines for convenience.

This dipeptide itself now has a COOH group on one end
and an $NH_2$ group on the other. Each of them can combine
with the appropriate group on the end of another amino
acid to lengthen the chain further, and in the same way
those new ends to the chain can make further links, making
a polypeptide. Just like the polymers described in chapter
five, the result is a zig-zag chain. But in this case the spine
of the long molecule is formed by alternating carbon and
nitrogen atoms (two carbons, one nitrogen, two carbons,

one nitrogen . . .). Starting at some arbitrary point in the chain, we have one carbon atom which has both a single hydrogen atom and the characteristic group of a particular amino acid attached to it as a side chain. This is bonded to a carbon atom which has an oxygen atom attached to it, and that carbon atom is bonded to a nitrogen atom, which in turn attaches to the next carbon atom that carries a complex side group. So the pattern repeats. The general structure is like this:*

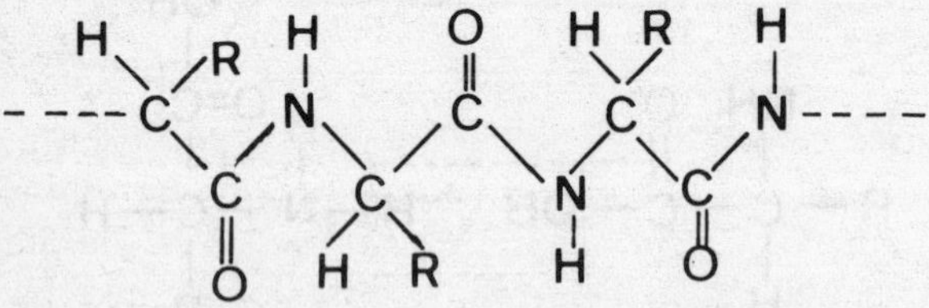

One of the key features of this chain is that the peptide bond forms a rigid structure, held firmly in place by quantum mechanical resonance. It was this that gave Pauling a clue to the way protein chains coil up. But we can bring out the most important feature of the protein chain by ignoring the zig-zags and looking just at the main spine with its side groups attached. The carbon atoms in

---

* The structure makes it clear why nitrogen is such an important ingredient in the molecules of life. Plants are able to take nitrogen from the environment (either from compounds in the soil, or directly from the air) and use it to make amino acids and proteins. Animals, people included, cannot do this. We depend on plants as the original producers of all the nitrogen-bearing molecules that we use. In the case of people, we can make 11 of the basic amino acids, but only provided that we have a diet which includes a sufficient quantity of nine essential amino acids, histidine, isoleucine, leucine, lysine, methionine, phenylalanine, threonine, tryptophan and valine. If just one of these is absent from our food then, eventually, we will die. And although we may get some of our essential amino acids from eating other animals, they in turn obtained them, originally, from plants, plants which fix nitrogen out of the environment and into organic compounds.

the main chain which are attached to the amino acid side
chains are labelled with the Greek letter alpha, α, to distinguish them from the carbon atoms which form part of
the peptide bonds between adjacent amino acid
components:

$$-N-C_\alpha-C-N-C_\alpha-C-N-C_\alpha-C-$$

Even more simply, in a purely schematic representation
taking no account of the geometry at all, we can indicate
the structure of the molecule by putting all of the amino
acid side chains on the same side of the spine:

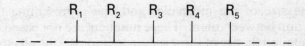

$$R_1 \quad R_2 \quad R_3 \quad R_4 \quad R_5$$

Displayed like this, it is easy to see how the chain carries a
message – the 'words' are the side chains, the residual bits
of the amino acids that make them different from one
another. It is no mystery that a long string of such 'words'
conveys a biologically important message. So, in the 1930s,
biologists wanted to know the order in which the amino
acids joined together to make different protein chains – far
from simple, but essentially a chemical problem. They also
wanted to know what shapes the chains folded into, their
three-dimensional structure, and why they should take up
particular shapes, if, indeed, it were true that all molecules
of a particular protein did form the same three-dimensional shape. The answers came from a combination of
more ideas about chemical bonding and from X-ray
diffraction studies, using X-rays to probe the structure of
protein molecules in crystalline form.

# Weak links can be important

The strong chemical bonds are not the only forces that operate between atoms, either between atoms on the loose, or members of a molecule, or between wandering atoms and molecules. All molecules exert a weak attraction, called the van der Waals force, on each other. In the 1870s, Johannes van der Waals, a Dutch physicist, improved the equations describing the behaviour of liquids and gases (the 'equations of state') by introducing terms which took account of the fact that molecules are not mathematical points, and that they are affected by attractive forces between each other, not just by perfectly elastic collisions. In 1881, he produced a new version of the so-called ideal gas law (the law which *does* assume that molecules are mathematical points involved in perfectly elastic collisions) which included two new constants, one for the size of the molecules and one representing the attraction between them. These numbers are not based on any detailed theory of the structure of atoms – the electron hadn't even been discovered then – but were chosen so as to fit the way real gases are known to behave. Eventually, by comparing the equations with the measured properties of real gases, and adjusting the equations accordingly, this was developed into a form that applied to all gases, and van der Waals received a Nobel Prize for the work in 1910. But a proper understanding of just what the van der Waals forces are had to await the picture of atoms and molecules developed from quantum physics.

The picture that emerges is simply that the electron cloud around a molecule is attracted by the positively charged nucleus of another atom, and vice versa. There is a compensating repulsion caused by the interaction of one negatively charged electron cloud with the other, and between the two positively charged nuclear regions, but until the molecules or atoms get close to one another this is not sufficient to completely outweigh the attractive force.

It was Fritz London, the Polish-born German physicist

who, together with Heitler, had developed the first quantum mechanical treatment of the hydrogen molecule, who explained the nature of the van der Waals force fully in the 1930s. Leaving aside the mathematical details, the reason for the force can be understood in terms of two spherical atoms, each with its own electron cloud, approaching one another. Although the net charge on each is zero, the electrons in each cloud are distributed around the central, positively charged nucleus. So the electrons, especially those in the outer shells, are relatively strongly influenced by the neighbouring atom's nucleus, while their own nucleus, buried in a sea of negative charge, is less influenced by the neighbour's nucleus. What the first atom 'sees', through the obscuring influence of half its own electron cloud, is another electron cloud. It is only when the atoms get so close together that the two electron clouds begin to interpenetrate that the two nuclei, no longer obscured from one another, repel each other sufficiently to stop the atoms getting closer together. The van der Waals force between atoms and molecules is stronger if there are more electrons in the screen, but it is only at all strong over a very short range of distances. In effect, the point of closest approach at which the repulsive and attractive electrostatic forces balance out can be called the radius of the atom or molecule – the van der Waals radius.

For non-ionic compounds, van der Waals attraction is the basis of the differences between solid, liquid and gas. In a solid, these forces hold the molecules of a particular compound together. When the solid is warmed, heat energy makes the molecules vibrate against one another, and when they gain enough energy they become unstuck and able to move past one another, although the van der Waals force still plucks at them as they pass. Provide still more energy in the form of heat, however, and the plucking becomes ever more ineffectual, until a point is reached where the fast moving molecules never come into contact long enough for the attraction to have much influence. Flying free, the molecules bounce off each other like hard spheres, in the gas phase. Because the van der Waals force is stronger for larger (heavier) molecules, by and large this

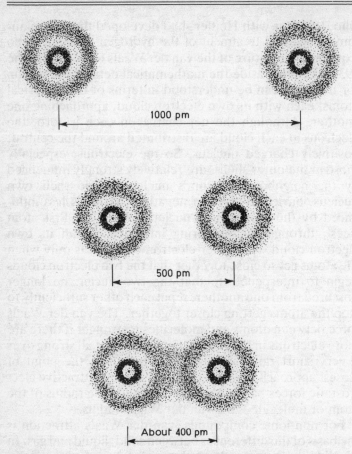

Figure 6.5 Two atoms attract one another as long as each positively charged nucleus 'sees' a negatively charged electron cloud beyond its own electrons. This is the basis of van der Waals attraction. Once the two electron clouds interpenetrate, however, the two nuclei 'see' each other, and the attraction is counterbalanced by the natural tendency of the two positive nuclei to repel each other. (A picometre is one millionth of a millionth, $10^{-12}$, of a metre.)

means that substances with larger molecular weight have higher melting points and boiling points than those with lower molecular weight. The chief exception is water, and you are about to find out why.

These forces, of both attraction and repulsion, are

important in long molecules with side chains, especially when the molecules fold up. Some parts of the chain will tend to stick together; other parts will tend to push away from each other. Overall, the molecule always tries to settle into the lowest energy state. Important though they are, however, the van der Waals forces are simply the fine tuning as far as the chemist trying to determine molecular structures is concerned. Where proteins are involved, the disulfide bridges that form so readily between two cysteine groups are much more important. If a long protein molecule, a polypeptide, doubles back on itself then such a link may easily form, holding it in a hairpin shape. And, of course, such bonds can equally readily form across the gap between two different polypeptide chains, holding two different protein molecules together:

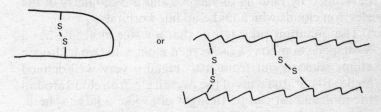

Finally, as far as the major influences on molecular structure are concerned, there is another kind of bond altogether, one which has its origins deeply rooted in quantum physics. This is the hydrogen bond.

# The hydrogen bond

Water is a most unusual liquid. It is made up of two hydrogen atoms and one oxygen atom, with a combined molecular weight of only 18 daltons. Yet it is a liquid at room temperature, while many very similar, or heavier, molecules are gases under everyday conditions. Carbon dioxide, for example, has a molecular weight of 44, hydrogen sulfide 34, methane 16 and nitrogen dioxide 46. According to the standard interpretation of the way van der Waals attraction operates between molecules, water has no right to be a liquid under the conditions that prevail on the surface of the Earth. And yet, not only is water a liquid, it is so far and away the single most vital liquid for us that 75 per cent of our body weight is in the form of molecules of $H_2O$. How can this be?

The answer must be that there is another force of attraction between water molecules, stronger than the van der Waals force, but weaker than the covalent bond which holds groups of atoms together as molecules. It is called the hydrogen bond, and, just as van der Waals' work only became explained by the ideas of quantum physics in the 1930s, so the puzzle of the remarkable affinity of water molecules for one another only became clear at about the same time, directly as a result of the understanding developed by Pauling of valence and the geometry of the electron clouds which make up filled orbitals.*

The quantum rules tell us that a water molecule has a V-shaped geometry, a fat oxygen atom with two hydrogen atoms sticking out from it to make a very well-defined angle, 104.5°. In terms of the overall electron cloud around the molecule, it can be thought of as like a large sphere, with two bulges swelling out from its sides (Figure 6.2); the surface of the overall lumpy shape represents the size of the molecule in terms of van der Waals force, the point at which attraction is balanced by repulsion. Whichever picture you prefer, however, the important point is that each hydrogen atom has only one electron to contribute to the makeup of the molecule. That electron, paired with an

---

* The earliest history of the idea of a special kind of bond between some molecules which contain hydrogen, especially the ammonium ion, goes back only to 1903 and the work of Alfred Werner. It was in 1920 that W. M. Latimer and W. H. Rodebush, of the University of California, Berkeley, suggested that the cause of the attraction between water molecules might be that 'a free pair of electrons on one water molecule might be able to exert sufficient force on a hydrogen held by a pair of electrons on another water molecule to bind the two molecules together' (*Journal of the American Chemical Society*, volume 42, page 1419). But an understanding of the physical basis for this behaviour came only in the 1930s.

electron from the oxygen atom, forms a cloud of negative charge which is most strongly concentrated in the region between the hydrogen nucleus and the rest of the oxygen atom. Although some of the electron cloud extends around the 'back' of the hydrogen nucleus, this nucleus, a single positively charged proton, is only thinly screened from the outside world. Overall, part of the electrically positive character of each of the two hydrogen nuclei shows through the electron cloud.

On the other side of the oxygen atom from the two hydrogen atoms, however, the situation is rather different. In effect, the oxygen has gained electrons from the two hydrogen atoms, and displays, from one side, the appearance of a full shell of eight outer electrons, wrapped around two inner electrons and a nucleus carrying only eight units of positive charge. From that side, with the hydrogen atoms concealed by the bulk of the oxygen atom, the molecule appears to carry an overall negative charge. Really, of course, taking the molecule as a whole it has zero net charge. But, equally really, that charge is unevenly distributed, with an excess of negative at the oxygen end, and two concentrations of positive at the hydrogen ends of the triangle. So when water molecules jostle against one another, there is a tendency for the hydrogen atoms on one molecule to link up with oxygen atoms on other molecules, positive to negative (Figure 6.6). Because the angle between the two bonds on each water molecule is very close to the angle between tetrahedral bonds (as in molecules such as methane, or the bonds between carbon atoms in a diamond), water molecules can form an array in which each oxygen atom is joined by hydrogen bonds to two other water molecules, and each hydrogen atom is joined in the same way to one oxygen atom as well as its molecular partner. This happens when the water is cold enough to solidify, and the resulting structure is a crystal of ice.

So ice is, in many ways, structurally similar to diamond. This structure is responsible for many interesting phenomena, including the beautiful geometry of snowflakes and the fact that solid ice occupies a greater volume than the

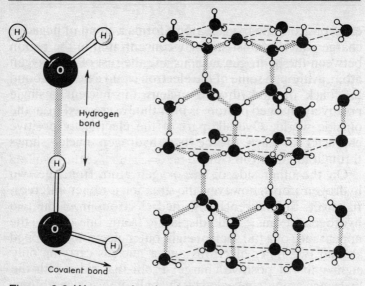

Figure 6.6 Water molecules have an affinity for one another through the hydrogen bond. A positively charged hydrogen nucleus which 'belongs' to one oxygen atom can still be attracted to the negatively charged electron cloud of a nearby oxygen atom in another water molecule. In ice, this attraction causes the molecules to form a crystalline array similar to the structure of diamond, but not as strong. This very open structure gives ice its low density, and explains why ice floats on water.

equivalent amount of liquid water, thanks to the wide spacing of this crystalline structure. That is why ice floats in water, or in your gin and tonic. Even when the ice is warmed so that the hydrogen bonds are broken and it melts into liquid water, the molecules brushing past each other still feel a strong attraction, much stronger than the ordinary van der Waals force, and that keeps them in the liquid state until the temperature reaches 100 °C.

Hydrogen itself is unique. It is the only atom which only has a single electron, and it is the only atom which can act as the positive partner in this kind of bond, or bridge, between molecules. It only does so when it is bonded to a suitable atom, but there are several other atoms, as well as oxygen, which can act as the negative partner in such liaisons. Hydrogen bonds never form with carbon atoms themselves, because of the orbital geometry and the rules of quantum physics. But they do form both with oxygen and that other important constituent of the molecules of life, nitrogen*, under the right conditions. The hydrogen atom may form a bridge between two oxygen atoms, as in water, or between two nitrogen atoms, or between one oxygen atom and one nitrogen atom. The right conditions might very well be as a bridge between two polypeptide chains, like the disulfide bridge, or, again like the disulfide bridge, as a link across a fold in a polypeptide chain, from one part of a protein molecule to another part of the same molecule. Hydrogen bonds are, indeed, common among biological molecules, and help provide the more complex ones with a very precise three-dimensional structure, the same for all molecules of the same compound.

# X-rays and proteins

There are three ways of approaching the problem of working out the real three-dimensional shapes of complex molecules of life. One is to probe their structure with beams of X-rays, study the way the X-rays diffract, and try to work out the structure from the diffraction evidence. Another is the traditional chemical approach, breaking the molecule up into its components, identifying them by chemical means and working out how they stick together. The third, which must draw heavily on both the other two approaches, is the approach of the theorist. Knowing the general rules which make atoms stick together in molecules, and understanding the nature and geometry of the different bonds in substances like amino acids, it ought to be possible to work out at least an approximation of the

---

* That is why ammonia, $NH_3$, also stays liquid at much higher temperatures than it 'ought' to, with a molecular weight of only 17. But the quantum geometry, with three hydrogen atoms attached to each nitrogen, doesn't allow for such effective hydrogen bonding between ammonia molecules as between water molecules, so ammonia is not such an interesting substance as water.

kind of structure a macromolecule, such as a protein molecule, ought to have. In fact, of course, all three approaches are mutually interdependent, and progress from the 1930s onwards came eventually through a fusion of all three lines of attack on the problem. But for a surprisingly long time – the best part of two decades – there was a fairly clear distinction between the approach based on chemical analysis and the approach based on X-ray studies of the molecules, while the first really dramatic breakthrough came from Pauling's highly theoretical approach. It would be confusing to try to indicate the way progress was being made on all fronts more or less simultaneously in the 1930s and 1940s, and it is much more logical to describe the advances made in each of the two experimental areas, in particular, separately. Which technique is discussed first is essentially an arbitrary choice; I have chosen to begin with the X-ray studies, as much as anything else because they provide a reminder of how deeply our modern biological ideas are rooted in the fertile soil of quantum physics.

Once X-rays had been identified as things that behaved, under the right sort of experimental conditions, like waves, the very techniques which had confirmed their wave nature could be turned around to probe the structure of matter. A pattern of atoms arranged in a crystal diffracts X-rays directed on to it, producing several diffracted beams that interfere with one another, strengthening the waves in some places and cancelling them out in others. Like the interference from two beams of light shining through two pinholes in a screen, the result is the X-ray equivalent of a pattern of dark and light stripes on a screen. But the X-rays used in this kind of work have a wavelength four thousand times shorter than the wavelength of visible light. And the details of the pattern depend on the arrangement of the atoms within the crystal, atoms which are spaced at distances very comparable to the wavelength of the X-radiation. Of course, X-rays are invisible. But they affect photographic plates and film, so it is a relatively simple matter to obtain pictures of these diffraction patterns. It is much harder, however, to interpret the pattern of dark and light spots and stripes in terms of the structure of the crystal being examined.

It is hardly surprising, in view of his involvement in the work that established the wave nature of X-rays, that Lawrence Bragg was the first person to use the technique to determine the structure of a crystal, in 1912. The substance he studied was common salt, NaCl, and the structure he found was of an array of alternating sodium and chlorine ions in a simple cubic lattice. Bragg invented X-ray crystallography. After his army service during the First World War, he returned to the theme, investigating more complicated substances which have correspondingly complicated structures. Meanwhile, among the new generation of chemists, Linus Pauling, on the other side of the Atlantic, began to learn about X-ray crystallography (initially from the book written by Bragg and his father), and completed his own first determination of a crystal structure in 1922 (the crystal was molybdenite). Working independently, but in parallel with each other, Bragg and Pauling both developed a set of rules for interpreting the X-ray diffraction patterns from more complicated mineral crystals; Pauling, reportedly to Bragg's chagrin,* published them first, in 1929. To this day they are known as 'Pauling's Rules'. This was the beginning of a friendly rivalry between Pauling and his colleagues, on the one hand, and Bragg's laboratory, on the other, that was to continue for decades, and which played some part in the search which led to the double helix.

As chemists – not just Pauling and Bragg, but many others as well – became more adept at using the technique, they turned their attention to ever more complicated molecular structures, and eventually, inevitably, to the molecules of life. Proteins come in two main varieties, the fibrous proteins in which the molecules retain, to a large extent, the long, thin, structure that you automatically associate with a chain, and the globular proteins, in which the chain is screwed up into a ball. The toy known as 'Rubik's snake' gives you a pretty good idea of the difference between the two. Like a polypeptide chain, the

---

* See, for example, Judson, page 76.

'snake' is made of rigid 'bonds' which are held together by joints which allow some rotation in certain directions. You can make the snake into a long zig-zag, or you can put various odd bends in it, and you can double it back on itself. You can even, if you remember just where to put the bends, curl it up into a ball. Fibrous proteins, the long thin ones, were the first proteins to be successfully investigated by a combination of X-ray diffraction pictures and inspired theorising.

# The alpha helix

The first X-ray diffraction pictures of fibrous protein came from William Astbury, a former student of the older Bragg, who was working at the University of Leeds in the early 1930s. His specific project was research into the physics of textiles, especially wool, which is largely made of the protein keratin. This is also found in hair and finger-nails. At that time, the X-ray diffraction photographs could not provide enough information to determine the structure of keratin. But they did show the kind of regular pattern which could only mean that there was a regular, repeating structure in the protein molecules – not as regular and ordered as a crystal of sodium chloride, but still clear evidence of order and repetition. In fact, Astbury found two different patterns, one characteristic of unstretched fibre, which he called the α-keratin pattern, and one typical of fibres that had been stretched, which he called the β-keratin pattern.

Astbury, and other researchers, suggested all kinds of structures for the protein chains which might account for the X-ray patterns, but none of them worked. Pauling tells how he 'spent the summer of 1937 in an effort to find a way of coiling a polypeptide chain in three dimensions, comparable with the X-ray data reported by Astbury'.* The

* Judson, op. cit.

effort was unsuccessful, and he decided that the way out of the impasse was to go back to first principles and take a good look at the structure of the building blocks of the polypeptide chain, the amino acids themselves and the bonds between them. Only then would he try once again to fit the pieces together to make a fibrous protein molecule which matched the X-ray data.

The investigation took a long time, even though Pauling was joined at CalTech by another researcher interested in the problem, Robert Corey. Together they studied X-ray diffraction photographs of single amino acids, dipeptides and tripeptides, using the information to probe the exact nature of the peptide bond. They found, among other things, that the C-N bond in the peptide link is shorter than it 'ought' to be, and that, thanks to a quantum resonance, it has a partially double-bonded character, like the resonant bonds in the benzene ring. This makes it impossible for the protein chain – the Rubik snake – to rotate around these bonds, and holds the whole peptide linkage flat, like this:

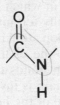

Both of the two bonds attached to the α-carbon atom, however, are free to rotate. It is as if the snake has two flexible linkages, then one rigid one, then two more flexible joints, and so on in a repeating pattern.

With another World War intervening, it took more than ten years before this painstaking research came to fruition. In 1948, however, Pauling knew he was on the right track. Simply by drawing out a representative polypeptide chain (like the one on page 152) on a long strip of paper, and bending it into a concertina shape to make the important carbon bonds have the right tetrahedral angle to one another, he could see how the whole chain could twist into

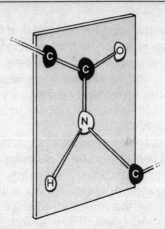

Figure 6.7 The peptide bond is a rigid structure. The C-N link is a hybrid intermediate between a single and a double bond, and no rotation can occur around it. This is a purely quantum physics phenomenon.

a helix, a corkscrew of repeating molecular linkages spiralling through space. Best of all, by getting all the angles just right he found a pattern in which the N-H groups of one peptide bond automatically fell into line with the oxygen atom attached to the carbon atom in the peptide bond four amino acids back down the chain, providing a perfect opportunity for hydrogen bonds to form. *Every* peptide bond in such a helical chain is involved in this hydrogen bonding, so that each coil of the helix is held to its neighbouring coils by several of these bonds. That is exactly what was needed to explain how the helix could hold its shape, without the whole molecule getting bent and distorted.

While Pauling was playing with his paper models, Bragg and his colleagues (by now, Bragg was the head of the Cavendish Laboratory, in Cambridge) were tackling the problem head on, using better X-ray diffraction pictures. They published their proposed solution for the structure of α-keratin in 1950. But it included one fatal flaw, allowing the structure free rotation about the peptide bond. This was, in truth, a chemical howler, for Pauling's work on resonant structures was very well known by then, and his

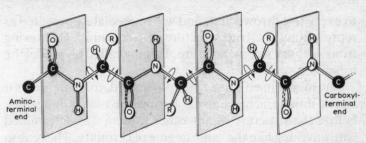

Figure 6.8 Because of the rigidity of the peptide link, a polypeptide chain can only fold up in certain ways. The combination of rigid links and allowed rotations enables it to coil up like a Rubik snake.

classic book *The Nature of the Chemical Bond* had been in print since 1939. Meanwhile, in 1949 and 1950 further X-ray studies in Pauling's lab confirmed the plausibility of the single stranded helix, held together by hydrogen bonds, as the basic structure of hair fibre. Pauling and his colleagues developed two variations on the theme, one less tightly wound than the original paper model, and published their detailed results in 1951. Following a couple of preliminary, throat clearing scientific papers, they astonished the world of biochemistry in 1951 with a remarkable tour de force, publishing seven separate scientific papers in the May issue of the *Proceedings of the National Academy of Sciences*, covering the structure of hair, feathers, muscles, silk, horn and other proteins, as well as spelling out details of the positions of the atoms in the two different helical models. The second version quickly fell by the wayside. But the original helix stood up

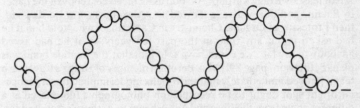

Figure 6.9 When a polypeptide chain coils up into a helix, the alpha helix, hydrogen bonds (dashes) help to hold the structure together.

to every test thrown at it, and was immediately accepted as representing the true structure of α-keratin.* Borrowing from Astbury's nomenclature, Pauling called his model the α-helix.

The model itself was a triumph; indeed, for many years it was a unique achievement, because it took a long time before the structure of any other protein was determined with anything like the same degree of certainty. The reason for Pauling's triumph, however, is as important for the story of the double helix as the fact of the triumph. It was Pauling's approach to the problem which was, within a couple of years, to encourage a new and even more important breakthrough. The fact that he found a helix certainly set people thinking even harder about helices than they had before; and at least some of those people he set thinking learned the lesson that one way to determine the structure of very large biomolecules is to start with the basic building blocks, and to try and find ways of arranging those building blocks in accordance with the known rules of chemistry, taking particular account of the ways hydrogen bonds might form to stabilise the overall structure. Instead of pulling the structure apart to see how the bits fit together, you start with the right bits and try to build a copy of the structure. As Pauling himself has emphasised, the α-helix structure was determined 'not by direct deduction from experimental observations on proteins but rather by theoretical considerations based on the study of simpler substances'.†

---

* Not least by Bragg's group. As soon as he received a copy of the paper by Pauling and Corey, Bragg showed it to Alexander (now Lord) Todd, then Professor of Organic Chemistry in Cambridge, who 'told him if he had asked me at any time in the past ten years, or if he had asked anybody in my lab, we could have told him that the peptide bond was planar'. (Judson, page 89.) Max Perutz, a member of Bragg's Cavendish team, noticed immediately that the α-helix model implied that a particular bright spot ought to be recorded on photographic film placed at a different angle from the fibre and X-ray beam to Astbury's standard set-up; with one photographic exposure in the right place, he found it straight away. For the second time, Pauling had scooped Bragg.
† *Chemistry*, page 496.

# Single and triple helices

Since 1951, a great deal more research has confirmed the nature of the helix often found in fibrous protein. In fact, it is rare for the helix to dominate the entire length of a protein chain. Folds and cross linkages alter the overall shape of many proteins, and there may be stretches dominated by the α-helix, as well as straight stretches and other portions held by different cross linkages even along the length of a single polypeptide chain. That wealth of biochemical information is, however, largely outside the scope of this story of the double helix. But it is surely worth a brief diversion to look at some of the ways in which simple linkages between α-helices themselves can explain some of the structure of our bodies, and of other animals, and why keratin can make substances which are superficially as different as hair and tortoise shell.

The geometry of the α-helix arises because the family of proteins that together make up the keratins contain amino acids that neatly fit such a structure, and do not, by and large, contain amino acids that would distort the helix. In addition, those helices contain a lot of cystine residues, the groups that are capable of forming disulfide bridges between polypeptides. In hard keratins, such as tortoise shell or horn, the helical coils lie side by side, joined to each other by many disulfide bridges, as indicated in Figure 6.10. These bridges are true covalent links, remember, much stronger than hydrogen bonds. So the result is a sheet of keratin molecules held firmly together. Add more sheets above and below, and it is easy to see how nature builds up the structure of a tortoise shell, or your own fingernails. What about your hair?

In fibres of hair, the strength of the bonding between individual α-helices is again the disulfide link. But in this case sets of three α-helices are twisted together, like the strands of rope, to make up a supercoiled triple helix. The disulfide bonds hold the three members of each triple helix tightly together, and the three helices all run in the same direction, in the sense that their amine groups are all at the

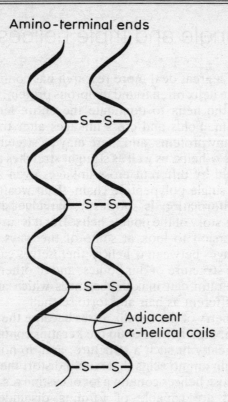

Figure 6.10 Alpha helices can also be joined to their neighbours, by disulfide bridges.

same end of the rope. Eleven of these three-stranded ropes are bundled together to make one hair microfibrile, and hundreds of microfibriles are bundled together to make one hair. Once again, even without going into details, it is straightforward to see how the properties of human hair are related to the sub-microscopic properties of the protein molecules.

It is also straightforward, on this picture, to understand something of the hairdresser's art. If hair is treated with a chemical compound that breaks up the disulfide bridges, the bonds between individual strands in the triple-helix ropes will be weakened. The hair will become soft and easily manipulated, and can, for example, easily be curled into a new shape. Then, when you have made the curls, all

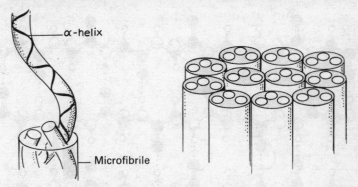

Figure 6.11 Another way for α-helices to get together is in the form of a triple-stranded 'rope'. This microfibrile is the basic constituent of human hair. The microfibriles pack together in large numbers to make a single hair.

you have to do is wash the hair with another chemical, one that attracts hydrogen out of the cystine residues and allows new disulfide bridges to form between the adjacent α-helices in their new configuration. Take the curlers away and the hair stays 'set' in its curly shape, thanks to the new disulfide linkages. Pauling and Corey's masterpiece of chemical deduction can, among other things, explain the 'permanent wave'. The phenomenon depends on the nature of the chemical bonds between sulphur and hydrogen atoms, and Pauling explained chemical bonding in terms of quantum physics. The 'permanent wave' is a phenomenon of quantum physics.

What of the β-keratin structure, distinguished by its unique X-ray fingerprint? That, it turns out, is not a helical structure at all. Instead of the polypeptide chains coiling around in helices, they form zig-zags very much more like the representation on page 152. And instead of the hydrogen bonds forming between different atoms on the same chain, they form, in much the same way, between the equivalent atoms in adjacent chains. The result is a structure which is superficially somewhat like the structure of a hard keratin, but in which the links between chains are formed by hydrogen bonds, not disulfide bridges. So the whole thing is much softer and more flexible – indeed, one fibre which has this pleated sheet structure, as it is called, is

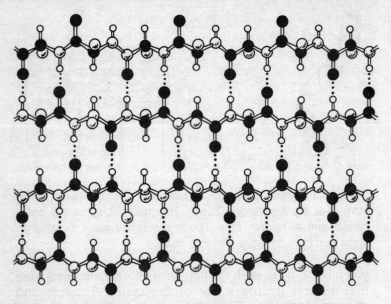

Figure 6.12 Another structure formed by polypeptides. Instead of coiling into helices, in the beta sheet the zig-zag chains of amino acids run side by side, held together by hydrogen bonds. The result is a soft strand – silk is made up like this.

silk. Armed with this kind of information, biochemists can now account for most of the structures in your body, the fibrous proteins which form its physical bulk. Collagen, for example, is the most common protein in the body. It is literally what holds you together, an important part of skin, tendons, cartilage and bone. And collagen, like hair, is made of three-stranded helical chains, different from the triple α-helices of hair but with a family resemblance. Triple helices are what holds you together; protein provides the scaffolding for the whole body. But where are the blueprints which define the structure, and where are the engineers who carry out the plans? Taking the second part of the question first, the construction engineers responsible for building the body and keeping it in shape can be identified with the other family of proteins, the globular proteins, and more specifically with the substances called enzymes.

# Enzymes

The first X-ray diffraction pictures of fibrous proteins and cellulose date from 1918; Astbury's pictures with their tantalising hints of regularity in the structure were obtained in the late 1920s and widely published in the early 1930s. Crystals of globular proteins were analysed in the same way for the first time in 1934. But it took two decades for molecular biologists to begin to come up with structures to account for the X-ray diffraction patterns of even the simpler globular proteins. The story is one that I only have space to touch on lightly here, but it is of the greatest importance in understanding how the body works. Indeed, the study of biochemistry is, in large measure, the study of enzymes – and enzymes are globular proteins.

Enzymes are the molecules in the body that encourage, or inhibit, chemical reactions between other molecules. In chemical terminology, they act as catalysts, altering the rate at which chemical reactions occur, without being altered by those reactions themselves. The best way to understand how this can happen is to think of a large, roughly spherical molecule – a globular protein – which has an indentation on its surface perfectly shaped to hold two specific and much smaller molecules. When those two different molecules sit in the cavity so conveniently provided by the enzyme, they will be held in alignment in just such a way that bonds can easily form between them. Instead of two molecules, we have one, and the enzyme can now release it to go about its biochemical business inside the cell, taking in two more small molecules (exactly the same as the first two) from the brew of chemicals around it and repeating the task as necessary. In a similar way, some enzymes break other molecules apart.

This is a gross oversimplification, but the image will do for our present purposes. Think of the enzymes, all of them globular proteins, as single-minded robots, each with its own specific task. One enzyme will join together one pair of molecules, a link in a polypeptide chain, perhaps, or the molecules involved in providing energy for your muscles. Another enzyme will be devoted solely to breaking one particular bond between one pair of organic molecules. In many ways, they resemble the idiot machine tools of a factory production line. How they go about their work is, as I have indicated, a story that has occupied many text-books, and, indeed, whole university degree courses. Let's just accept that they do work, and that their work depends very much on their structure. The relevant, and interesting, question for the story of the double helix is how they are themselves manufactured, and what determines the unique structure of each globular protein.

# Protein crystallography

The first X-ray diffraction photographs of protein crystals (as distinct from fibres) were obtained by J. D. Bernal and Dorothy Crowfoot, in Cambridge. Previous attempts had all failed because, it turned out, the crystals had been dried before examination; Bernal and Crowfoot showed that you could only get good X-ray diffraction patterns from proteins if the crystals were kept wet. In a contribution to a volume published to commemorate Pauling's 65th birthday,* Dorothy Crowfoot Hodgkin, as she then was, and Dennis Riley recalled those early days. John Philpott, a crystallographer working in Uppsala in the mid-1930s, had prepared crystals of the protein pepsin, by growing them in a liquid rich in the protein – this is the standard technique for preparing crystals, from their 'mother liquor', a concentrated solution. Philpott left his crystals in the refrigerator while he went on a skiing holiday, and was astonished to see how much they had grown on his return. He proudly showed the 2 mm long crystals to a visitor, Glen Millikan, who said 'I know a man in Cambridge who would give his eyes for those crystals.' Philpott gave him some, still

---

* *Structural Chemistry and Molecular Biology*, edited by Alexander Rich and Norman Davidson, W. H. Freeman, San Francisco, 1968. In this same volume, incidentally, J. D. Bernal himself describes Pauling's work on the α-helix as 'his greatest triumph' (page 270). High praise, when you recall that Pauling received two Nobel Prizes for his *other* contributions.

in the tube in which they were growing in the mother liquor, and so they travelled to Bernal's lab in Cambridge, in Millikan's coat pocket. Thanks to this mode of transport, Bernal was able to make the key discovery that the X-ray diffraction pattern only appeared when the *wet* crystals were analysed. The wet crystals have an ordered structure, but when they are dried the structure collapses, like a house of cards, into disorder. The first X-ray diffraction photographs of single pepsin crystals were obtained in 1934.

Bernal immediately appreciated the potential for the future if only photographs like these could be interpreted in terms of structure. But the problem he was faced with was far more difficult than the problem of the structure of fibrous proteins, which was itself 17 years from solution at that time. With hindsight, we can see why. The long polypeptide chain which we can write down on the page in chemical shorthand represents what is called the primary structure of a protein, the order of amino acids along the chain. The $\alpha$-helix, Pauling's brainchild, represents the secondary structure, the way in which the chain is coiled. But the structure of a globular protein depends on the way the *helix* itself is bent in three dimensions – it is, in the same nomenclature, a tertiary structure. It could only really be solved after the secondary, helical structure was itself understood. But that didn't stop the crystallographers from building up their store of information about globular proteins in the 1930s and 1940s.

Skipping rather lightly over a great deal of work by many accomplished scientists, the next significant landmark can now be seen as the arrival of an Austrian, Max Perutz, to work in Cambridge in 1936. After some grounding in the basics of crystallography, in 1937 Perutz acquired some crystals of hemoglobin, and began the work that was to lead him to the Nobel Prize. Why hemoglobin? Well, why not?: 'at that time the great unsolved problem was the structure of *proteins*. It didn't much matter which one you chose – or so it seemed. You wanted to study the structure of a protein.'[*] In 1938, Bernal moved to London, while

---

* Judson, op. cit.

Lawrence Bragg came to Cambridge to succeed Rutherford as the head of the Cavendish lab. Perutz stayed in
Cambridge, and soon fired Bragg, the crystallographer,
with his enthusiasm for proteins. Bragg soon obtained a
research post for Perutz, giving him enough security to
bring his parents over to England, and obtained funds for a
new X-ray tube. By 1940, the attack on the structure of
hemoglobin was in full swing. But progress was very slow.

Skipping on again, the next key development occurred
in 1946, when John Kendrew joined the team. Kendrew
was originally trained in physical chemistry, but his career,
like so many others, had been interrupted by war service,
during which he met Bernal, who at the time was scientific
adviser to Lord Mountbatten, the supreme allied com-
mander in Southeast Asia. Intrigued by the possibilities of
research in molecular biology, Kendrew returned to
Cambridge after the war seeking a place to work for his
doctorate. Perutz fixed him up as a research student, under
the formal supervision of Bragg, and set him to work on the
hemoglobin problem. In 1947, Bragg obtained support
from the Medical Research Council which established
Perutz and Kendrew as a two-man unit devoted to the task
of studying hemoglobin by X-ray methods.* But Kendrew
wanted a protein of his own to study, and it was
increasingly clear that hemoglobin was too large and complex a structure to be cracked easily, even then. Kendrew
picked myoglobin, a small protein very much like hemoglobin in many ways.

# Myoglobin and hemoglobin

Both myoglobin and hemoglobin contain atoms of iron,
locked up in a chemical package called the heme unit (see
Figure 6.1). Hemoglobin has four such units, all involved

---

* This became the MRC's now world famous Laboratory for Molecular Biology, with Perutz as its Director and Kendrew his deputy.

in holding on to and releasing oxygen, which makes it very good at its job of carrying oxygen from the lungs around the body in the blood stream. Myoglobin has only one heme group in each molecule, containing just one iron atom. It is about a quarter the weight of hemoglobin. Whereas hemoglobin carries oxygen in the blood, myoglobin is a muscle protein, and has the job of holding oxygen, brought to the muscle by hemoglobin, until the muscle needs it. Hemoglobin, thanks to its heme group, is what gives blood its red colour; myoglobin is what gives muscle its red colour.

Cutting a long story short, the attack on myoglobin first succeeded where the attack on hemoglobin had failed, and then provided the key to unlock the secret of the structure of hemoglobin itself. The breakthrough came from a technique developed by Perutz and Kendrew, which had its roots in Bragg's pioneering work on salt crystals. The idea is to replace some of the atoms, or ions, in the crystal by heavier atoms that affect the X-rays differently. In the salt example, Bragg compared the X-ray diffraction patterns of sodium chloride with those of potassium chloride; the differences told him something about the location of the sodium and potassium ions in the two crystals, while the similarities, obviously, could be related to the chloride ions, which were the same in both cases. In the early 1950s, in the midst of their painstaking work on hemoglobin and myoglobin, the team learned of the possibility of attaching a mercury atom to the chain, using its great affinity for the sulphur atoms found at the ends of molecules, or side chains, of cysteine. This jumped off from work by Austen Riggs, at Harvard, who was studying the chemical properties of hemoglobin. But Perutz and Kendrew realised that with a mercury atom, containing 80 electrons, attached to a chain of myoglobin or hemoglobin they would have a new tool with which to tackle the X-ray diffraction photographs. The heavy atoms themselves alter the diffraction pattern, in a way which depends on the structure of the crystal. By first determining the positions of the heavy atoms from the photographs, and then comparing the patterns produced with and without the heavy atoms attached, they were able, at last, to produce something like a contour map of the distribution of electrons within the crystal.

This first step was taken in 1953. In 1954, following a casual remark by Dorothy Hodgkin, the MRC team extended the work by finding (with very great difficulty) ways to attach other heavy atoms to the molecules, giving them another angle on the diffraction patterns that was eventually to lead to a full three-dimensional reconstruction. Kendrew, by this time, had focussed his own personal project on to myoglobin obtained from whale meat, which turned out to provide excellent crystals for his needs. By 1955, the team had managed to attach heavy atoms to myoglobin molecules at five separate sites; by 1957, they had determined the structure, and it was published, by Kendrew and five collaborators, in 1958 – the first protein whose structure had been completely determined. It had been a long, but worthwhile haul. The structure of hemoglobin followed in 1959, and the Nobel Prize for Kendrew and Perutz in 1962 (of which more later). The big surprise in all this, however, was the detailed similarity of the whale myoglobin to the hemoglobin they were working with, which happened to be from horse blood.

## A common basis

Perutz himself has summed up the findings in his book *Proteins and Nucleic Acids*, while Kendrew's slightly less technical account appeared as *The Thread of Life*. Between them, they provide as good a summary as you could wish of the determination of the structure of these important proteins. The complete X-ray pattern of a myoglobin crystal has something like 50 000 spots, which gives you some idea of the complexity of the molecule and the difficulty of determining its structure. Yet myoglobin is a relatively small protein. It contains 'only' some 2 500 atoms, arranged in a single polypeptide chain made up from 153 amino acid residues. The chain itself consists of seven straight stretches, each of them $\alpha$-helices, with bends between them. The helices take up 110 of the amino acid residues, and the corners, or irregular regions, are occupied by the other 43, a bit less than one third of the chain. All of these folds make up a little hollow, a sort of pocket in which one heme group nestles, like a marble nestling in the palm of your hand.

Hemoglobin is nearly four times as large. It is made of four polypeptide chains, which come in pairs (two identical alpha chains, as they are called, and two identical beta chains) but which are all very similar to the single chain of myoglobin. Both the alpha and the beta chains of hemoglobin have straight stretches of $\alpha$-helix which make up more than 70 per cent of their length; both have bends very similar to those in the myoglobin chain; and all these chains have very similar amino acids distributed in much the same way. The four chains in hemoglobin lock together, presumably held in place by electrostatic forces, to make a roughly spherical ball, with four pockets for heme groups on its surface (think of your fingers, interlaced to hold your two hands together). Roughly speaking, the sphere of influence – the van der Waals radius – of each atom along the protein chain is, indeed, spherical. Each overlaps with its neighbour – there is no empty space between the electron clouds. So a protein chain can be imagined as like a string of fat beads. The overall structure of a set of folded protein chains, such as a hemoglobin molecule, is very knobbly, with all the electron clouds packed closely together. The only gaps in the structure are the pockets for the working parts that allow the molecules to perform their biological functions, and the whole thing can be thought of, if you wish to have a mental picture, as looking rather like a ball of frog spawn, or a bunch of grapes. But the grapes don't grow out from a central stem, they are in the form of long chains folded back upon themselves and each other. The structural determinations were triumphs in themselves, even though the story is not directly part of the story of the double helix. But the revelation of the similarities between proteins that do the same job in different species, and similar jobs in the body of the same individual, is very much relevant to my main story.

It isn't just that horse hemoglobin is like horse myoglobin, or that horse myoglobin is like whale myoglobin, but that *horse hemo*globin is like *whale myo*globin. Further research since the 1960s has now identified the detailed structures of many other molecules of life, and always the story is the same. Evolution, it seems, is very conservative. Once a molecule has evolved to do a certain job it may be improved by natural selection, but it is not replaced by a completely different molecule.

The picture that emerges is like this. Some time ago, in the distant past history of life on Earth, a molecule arose which had the capability of binding to a single heme group and thereby transporting and storing oxygen. That prototype oxygen carrier must have been a single polypeptide chain, which over many, many generations of evolution and selection became more complex and more efficient, developing into the myoglobin molecule with its neatly folded shape, ideally designed to cup the heme group in place. Although many different species have evolved on Earth during that time, we all carry the descendants of those original oxygen-bearing polypeptide chains in our bodies. As each species has split off to form a new branch on the evolutionary tree and gone its separate way, so there has been an opportunity for minor differences in the proto-myoglobins to arise. But the family relationship, and common ancestral origin, can clearly be seen in the proteins of which we are made.

The investigation of hemoglobin establishes the picture and develops it further. A molecule capable of transporting more oxygen than a single strand of myoglobin did not arise independently, but clearly as a result of combinations of the original, proto-myoglobins. Perhaps this happened in two stages, with a double myoglobin-like molecule arising first, and later combining with another double molecule to make proto-hemoglobin. That would certainly fit the discovery that the hemoglobin molecule contains four chains which come in two varieties. But the details do not matter so much as the discovery that all of these oxygen bearing molecules, in animals as different from one another superficially as horse and whale, are clearly members of the same family, molecules descended from the same original oxygen carrier, back in the primordial ooze. What's more – although this really is getting ahead of the story, and the details will have to wait for Chapter Ten – the degree of difference between the hemoglobin of a horse and a whale gives us what turns out to be a very accurate indication of exactly how many millions of years have elapsed since the ancestral lines of the horse and whale split off from some common ancestor, an ancestor whose myoglobin and hemoglobin they inherited. All of that work depends on the one aspect of the determination of protein structure that I have left till last – the chemical analysis which reveals the exact order of the amino acid residues along the chain.

# Reading the message

Until 1942, there was no prospect of determining the amino acid components which made up even a simple protein molecule, and no way of finding out the order in which the amino acids (strictly speaking, the amino acid residues) were strung along the chain. This, of course, is one reason why the X-ray crystallography techniques were favoured by investigators in the 1930s. But the situation changed when the Englishmen Archer Martin and Richard Synge developed a refinement of the technique called chromatography, one of the standard tools in the chemists' armoury.

Chromatography depends on the physical properties of different molecules, especially the ease – or difficulty – with which they come out from a solution and stick to other molecules. There are various different techniques of chromatography, but the line developed by Martin and Synge is the one that is of the greatest direct relevance to the story of the double helix, and the one I shall concentrate on here. The approach stems from the very simplest kind of paper chromatography, a phenomenon known to every school-child who has ever dipped a piece of blotting paper into a puddle of ink, pulled it out and watched the way the ink soaks up through the paper. Ink is a mixture of dyes dissolved in water; it is the water which, by capillary action, climbs up the strip of paper and carries the dye molecules along with it. But the dye molecules come in different shapes and sizes, and some of them stay in solution better, while others tend to stick to the blotting paper and get left behind. The result is that as the ink rises through the paper different dyes move at different rates, appearing as distinct bands of colour, one behind the other.

A very similar technique can be used to separate a mixture of amino acids, or other complex biomolecules, from a solution. A spot of the solution is placed on one end of a strip of filter paper, and the paper dipped into a solvent which rises up past the spot, carrying the different amino acids, say, up through the paper at different rates, with the least sticky molecules moving fastest and travelling furthest before the paper dries.* When the paper is dried, the original brew of amino acids has been spread out into bands, one behind the other, but overlapping rather too much for detailed analysis to be effective. What Martin and Synge did was to find a way of separating the bands out still further, into distinct spots that each correspond to only one compound, one amino acid in the case of protein analysis. The trick is disarmingly simple in principle, though it requires care and skill to be useful in practice. After you have carried out a chromatographic separation of this kind, turn the piece of paper through a right angle, and dip the edge that is now at the bottom into another preparation of solvent – perhaps a different solvent from the one used first time. Once again, the solvent rises up through the paper, spreading the amino acid blobs out again in a direction at right angles to the original separation, and revealing much more detail as the large blobs get subdivided. All you have to do then is identify each of the blobs by standard chemical techniques (easier said than done, but possible), and you know exactly which chemical compounds – which amino acids – were present in the original spot of solution placed on the paper.

A variation on the theme involves attaching the strip of paper across a pair of electrodes, one positive and one negative. Positively charged fragments in the solution will

* It isn't necessarily the smallest molecules that are least sticky. What matters is the type and number of groups on the outside of the molecule, and whether these easily 'catch hold' of the solid material that the amino acids are sliding past, or not.

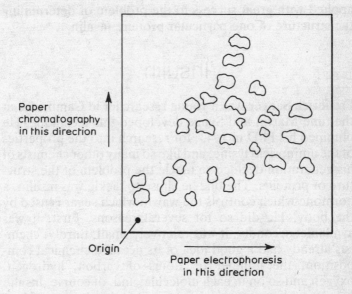

Paper
chromatography
in this direction

Origin

Paper electrophoresis
in this direction

Figure 6.13 When a biomolecule such as human hemoglobin is broken up into its constituent amino acids, the components can be separated by a combination of chromatography and electrophoresis, as explained in the text. Each blob is made up of molecules of one amino acid, and by identifying each component biochemists begin to understand the structure of the biomolecule of which the amino acids are components.

move towards the negative electrode, while negatively charged fragments move towards the positive electrode, and electrically neutral fragments stay put. Combine this technique, called electrophoresis, with ordinary chromatography at right angles to the electrical separation and, once again, you have a two-dimensional separation which can be used to pick out every constituent amino acid as a separate blob on the paper. It is possible, in principle, to read the amino acid message carried by the molecules of a specific protein. Martin and Synge received the Nobel Prize in chemistry 'for their discovery of partition chromatography' in 1952. By that time, the technique had been applied with great success to the problem of determining the structure of one particular protein, insulin.

# Insulin

Frederick Sanger was a young researcher in Cambridge at the time Martin and Synge developed this technique. He obtained his PhD in 1943, for research into the properties of the amino acid lysine, and like so many other chemists of his generation decided to tackle the problem of the structure of proteins. The one he chose to tackle was insulin, a hormone which controls the way in which sugar is used by the body. He did so for several reasons. First, it was available; secondly, it was relatively small; thirdly, chemists already had a good idea of its detailed chemical composition, the number of atoms of carbon, hydrogen, oxygen and so on in each molecule; and, of course, insulin has immediate practical importance in the control of diabetes.

Sanger worked with beef insulin. Each molecule of this contains 777 atoms, made up of 254 carbon, 377 hydrogen, 65 nitrogen, 75 oxygen and 6 sulfur. The problem he had to tackle, however, started from the whole molecules, which had to be broken up into pieces that could be analysed so that, with a knowledge of the constituent parts of the molecule, Sanger and his colleagues could find how the pieces of the jigsaw puzzle – the amino acids – were put together.

One of his key developments was finding a way to label the ends of the individual polypeptide chains in any protein molecule. A substance called dinitrofluorobenzene has two very valuable properties for this purpose. First, it attaches very strongly to the amino group on the end of an amino acid. In addition, it gives the amino acid it attaches to a bright yellow colour. Because the amino acid residues in a polypeptide chain are joined to one another by the peptide bond, each protein chain has only one amino group left intact, at one end of the chain. So when dinitrofluorobenzene is combined with the polypeptide, it attaches only at the end. Then, when the polypeptide is broken up into its component amino acids by hydrolysis, just one of these will carry the yellow marker. Starting with a protein molecule, which may contain a few chains, this labelling and splitting technique produces a brew of amino acids in which only the ones that used to be on the end of a chain are coloured yellow. Partition chromatography then spreads the amino acids out into spots on a filter paper, and the number of yellow spots tells you how many chains you have to work with.* What's more, by analysing the yellow spots, you can find out which amino acids are on the end of each chain.

Now things get complicated. With the aid of chromatography, Sanger and his colleagues found out exactly which amino acids were combined in the chains making up insulin, and they found that there were two chains in each molecule. Once the disulfide bonds holding them together were broken, the two chains could be separated from one another, using techniques which depend on the different size and molecular weight of the two chains. To find out the order of the amino acids along each chain, the team had to snip them apart more selectively, using enzymes that only break down the chemical bonds between specific amino acids, or using partial hydrolysis to break the chain into fragments, dividing it only at its weakest links, to produce shorter stretches of chain each made up of several amino acids. One of the chains they had to tackle contained 30 amino acids, the other 21; these were far and away the biggest polypeptide chains ever analysed at the time, although now much bigger chains have been analysed in

---

* In reality, of course, even a small spot of solution contains hundreds of thousands of protein molecules. But each coloured amino acid from the end of a particular polypeptide chain is identical, and they all move through the filter paper together to make a yellow spot containing hundreds of thousands of identical amino acid molecules. Because every molecule of the same protein is identical to every other molecule of the same protein, and made of identical sets of chains, the technique works just as if we were dealing with one individual molecule. And it *only* works because all the molecules of a particular protein are the same.

similar detail. One chain, the treatment showed, began with the amino acid glycine. When the chain was only partly broken up, it produced pieces that could still be separated by chromatography, and then each of those pieces, after separation, could be broken up completely into its constituent amino acids and analysed again by chromatographic techniques. In this way, Sanger's team found that the labelled broken ends of the chain sometimes produced glycine and isoleucine, while at other times the chain broke to give an end piece containing glycine, isoleucine and valine, while still other end fragments contained either the mixture glycine, isoleucine, valine and glutamic acid, or the same four amino acids plus another molecule of glutamic acid. This had to mean that the end of this particular polypeptide chain began:

Gly.Ileu.Val.Glu.Glu. . . .

In similar fashion, the painstaking analysis revealed the structure of other bits of chain from the rest of the molecule. After a year's work, the team had an unambiguous structure for one chain, a unique combination of amino acids that could explain all of the fragments found by breaking the chain into pieces. It took another year to analyse the other chain in the same way, and by March 1953 Sanger and his colleagues had published their detailed analyses of each of the insulin chains. Further work was necessary before it became clear exactly how the two chains are held together by disulfide bridges, but we now know that the structural formula of an insulin molecule can be represented as shown in Figure 6.14. Sanger received the Nobel Prize in chemistry for this work in 1958, and many proteins have been 'sequenced' in this way since.*

---

* One of the strangest quirks in the history of the Nobel Prizes concerns the date Sanger was honoured in this way. He developed the technique for analysing proteins in terms of their constituent amino acids, but the technique was very quickly picked up by other chemists, including the American Vincent du Vigneaud. Du Vigneaud used Sanger's technique to determine the structure of two, much simpler proteins, called oxytocin and vasopressin, and then went one better than

The fundamental importance of this work for the story of life, however, is not that it opened the way for biochemists to understand enzymes, hormones and the other molecules of life.

Sanger's work established beyond any remaining shadow of a doubt that proteins are composed of polypeptide chains, and that every molecule of a particular protein is made up of identical chains in which the same amino acids are arranged in the same order. The structure of each chain is *not* given by some simple chemical rule, such as 'glycine is always next to valine', or by simple repetitions, such as six leucines followed by four valines and two cystines, then repeat the whole pattern to the end of the chain. It really *is* best described in terms of a coded message. Nobel laureate Jacques Monod, speaking to Horace Judson, stressed the crucial importance of this work. 'Sanger's discovery,' he said,* 'revealed a sequence that had no rule.' And yet it contained information, and 'to explain the presence of all that information in the protein you absolutely needed the code'. Every protein molecule, in other words, contains a coded message which ensures that each globular protein has a specific shape which uniquely fits it to carry out its role as a molecule of life. Sanger provided the means for biochemists to read that message. But he did not solve the code itself. Where in the living cell is the essential blueprint which tells the cell how,

---

Sanger by constructing exact copies of these rather simple proteins by adding amino acids to one another in the right order. He synthesised, from the non-living building blocks of life, molecules of life that behaved in every way exactly like those produced in the body. The achievement was so dramatic, with all that it implies for our understanding that living things are no more than the sum of non-living molecules put together in the right order, that du Vigneaud received the Nobel Prize almost immediately, in 1955. We can imagine the light dawning on the Nobel committee a couple of years later when they realised that without Sanger, du Vigneaud could never have carried out his work, and that determining the structure of insulin was in any case a much knottier problem than determining the structures of oxytocin and vasopressin. Still, in the end justice was not only done, but seen to be done.

* *The Eighth Day of Creation*, page 213.

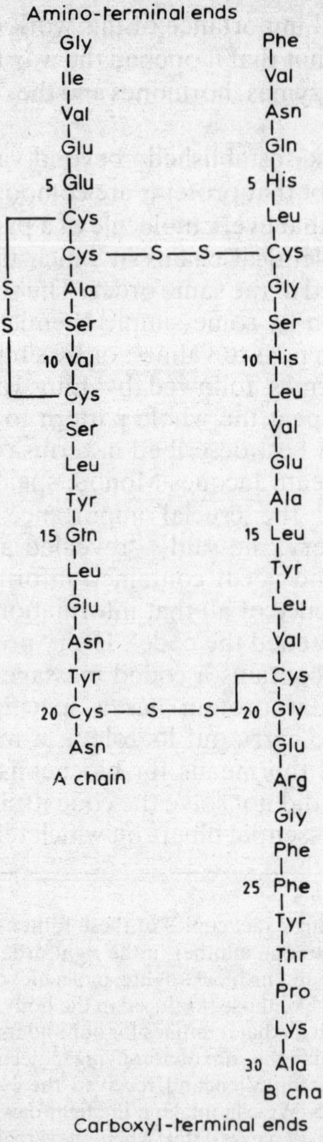

Amino-terminal ends

| | |
|---|---|
| Gly | Phe |
| Ile | Val |
| Val | Asn |
| Glu | Gln |
| 5 Glu | 5 His |
| Cys | Leu |
| Cys —S—S— Cys | |
| Ala | Gly |
| Ser | Ser |
| 10 Val | 10 His |
| Cys | Leu |
| Ser | Val |
| Leu | Glu |
| Tyr | Ala |
| 15 Gln | 15 Leu |
| Leu | Tyr |
| Glu | Leu |
| Asn | Val |
| Tyr | Cys |
| 20 Cys —S—S— 20 Gly | |
| Asn | Glu |
| A chain | Arg |
| | Gly |
| | Phe |
| | 25 Phe |
| | Tyr |
| | Thr |
| | Pro |
| | Lys |
| | 30 Ala |
| | B chain |

Carboxyl-terminal ends

Figure 6.14 Beef insulin was the first protein whose structure was completely determined. It consists of two polypeptide chains held together by disulfide bridges. The work earned Fred Sanger a Nobel Prize.

and when, to manufacture each kind of protein? And how do the cell's engineers – its own enzymes – carry out the task of translating that code into messages in the form of insulin, hemoglobin and the rest? The complete description of the second insulin chain was published in March 1953. Just two months later, Francis Crick and James Watson published their famous description of the double helix of DNA, the life molecule itself, and thereby pointed the way to a resolution of both these puzzles.

# CHAPTER SEVEN

# THE LIFE MOLECULE

The four main categories of biochemically important substances are the fats, sugars and starches (which are both polysaccharides), proteins and nucleic acids. The nucleic acids were the last of these to be identified, and their central importance to the story of life was not fully appreciated until the 1950s. But by then they had been known for nearly a hundred years, since the pioneering work of the Swiss biochemist Friedrich Miescher. DNA itself was actually discovered in the same decade, the 1860s, that Gregor Mendel published the results of his experiments with peas.

Miescher was born in Basel in 1844. His father was an eminent medical man, Professor of Anatomy and Physiology in Basel from 1837 to 1844, and young Friedrich decided to follow the same profession after his father had turned down an early proposal from the boy that he might enter the priesthood. Miescher's background was also medical on his mother's side. His maternal uncle was Wilhelm His, a pioneering researcher in embryology and the study of different body tissues, who also became Professor of Anatomy and Physiology in Basel – the same Chair Miescher's father had held – from 1857 to 1872. Friedrich Miescher was strongly influenced by his uncle, and it was largely thanks to the advice of His that when Miescher completed his medical training in 1868 he did not go into medical practice but went on to research into the chemistry of the cell. His himself had become aware, through his own research on tissue development, of the crucial role played by cell chemistry, and told Miescher that 'the final questions in the development of tissue could only be solved on a chemical foundation'.*

# Pus provides the clue

So Miescher moved from Göttingen, where he had been a medical student, to Tübingen, where he studied organic chemistry and began his research, in the laboratory of physiological chemistry founded at the university there by Felix Hoppe-Seyler. This was the first laboratory in the world devoted to what we now call biochemistry, and Miescher arrived at a time of great excitement. The theory that living organisms could arise from non-living matter by 'spontaneous generation' had only just been overturned, and it was during the 1860s that Rudolf Virchow developed the idea that living cells are created only by other living cells, while in 1866 Ernst Haeckel had suggested that the cell nucleus might contain all of the 'factors' necessary for the transmission of hereditary information. Proteins were already known to be the most important structural substances in the body, and Miescher set himself the task, encouraged by Hoppe-Seyler, of identifying the proteins present in some of the simplest human cells. The cells were the white blood cells present in large quantities in human pus, which Miescher could obtain conveniently from a nearby surgical clinic in Tübingen.

Miescher was looking for the most fundamental constituents of life, and he never knew just how well he succeeded. Odd though his choice of working material may seem today, in the 1860s, before the days of widespread use of antiseptics, almost any post-operative wound was likely to suppurate and produce pus, and by extracting this from the bandages given to him by the clinic Miescher could

---

* Quoted by Franklin Portugal and Jack Cohen, *A Century of DNA*, page 9.

obtain human cells to study without having to ask people for samples of blood or flesh, which might have proved rather difficult. His problems didn't end with the unpleasant nature of his working material, though. He had to find ways of washing the pus cells free from the bandages without destroying them, and then find chemical means of breaking into the cells and analysing their contents. In the course of these experiments, Miescher found that the cells he was working with contained a substance unlike any of the familiar proteins. He had found something new.

At first, the new cell chemical was thought to be another protein, but the chemical tests soon showed that it had a different chemical composition. The new substance appeared only when the cells were treated with a weakly alkaline solution, and by looking at the cells under the microscope Miescher noticed that the weak alkaline solution caused the nucleus of the cell to swell and burst open. He deduced that the new substance came from the nucleus itself, not from the surrounding protoplasm of the cell, and he developed techniques to separate the whole nuclei from the rest of the cells so that he could test this hypothesis. By the summer of 1869, Miescher had confirmed that the new substance came from the cell nuclei, and he had found the same material in the cells of pus, yeast, kidney, red blood cells and other tissues. He called the substance nuclein, because of its association with the cell nucleus, and he set about determining its chemical composition, making the key discovery that it contains phosphorus, in addition to the carbon, hydrogen, oxygen and nitrogen familiar from analyses of other molecules of life.

In the autumn of 1869 Miescher left Tübingen to work in Leipzig, but before the end of the year he had sent the manuscript announcing these important discoveries to Hoppe-Seyler, for publication in the Professor's somewhat immodestly titled *Hoppe-Seyler's Journal of Medical Chemistry*. A chapter of accidents repeatedly delayed the publication of the paper. First, Hoppe-Seyler was suspicious about the discovery, and refused to accept Miescher's work until he had carried out similar experiments and proved to his own satisfaction the reality of this new biochemical compound, nuclein. Then, the outbreak of the Franco-Prussian war in July 1870 made communications difficult (Miescher had by now returned to Basel) and interfered with the publication of any non-essential journals in what was then the German Empire. Finally, Hoppe-Seyler seems to have done the dirty on Miescher by holding up publication of the original paper announcing the discovery of nuclein still further, until two new papers reporting more work on the substance by two of Hoppe-Seyler's students could appear alongside Miescher's paper and Hoppe-Seyler's confirmatory paper in the same issue of the *Journal*. Perhaps Miescher should have taken more careful note of the journal's full title! Still, the paper was published, in 1871, and there has never been any doubt at all about Miescher's priority as the pioneering discoverer of what we now call nucleic acid.

## What is nuclein?

There was, however, considerable doubt and debate about the importance of nuclein, or nucleic acid, in the century that followed. Miescher's own work – back in Basel – concentrated on a long and detailed study of sperm cells from salmon. These proved ideal for his purposes. In any sperm cell the nucleus is extremely large (we now know that the sperm's sole purpose is to carry genetic material to the egg) and in salmon sperm the nucleus provides more than 90 per cent of the cell's mass. On their epic journey upriver to the spawning grounds, salmon do not eat, and grow thin as muscle tissues are absorbed into the body. Yet their store of sperm builds up to the enormous quantities required for successful spawning, and Miescher pointed out that the structural proteins of the body must be converted into sperm, an early realisation of the fact that different parts of the body can be broken down and rebuilt in another form. He found that nuclein was a large molecule which included several acidic groups (the term nucleic acid was introduced in 1899 by Richard Altmann, one of Miescher's pupils) and that it is associated in the nucleus with another substance, which he called protamine, that we now know to be a protein. In 1872, His moved from Basel to the University of Leipzig, and on the strong recommendation of both His and Hoppe-Seyler, Miescher was appointed Professor of Physiology in his place, with the title Professor of Anatomy being split off and a new Chair created. So Miescher followed both his father and his uncle into the same Chair, where he remained until his early death, of tuberculosis, in 1895 at the age of 51. His fatal illness was probably aggravated by the long hours he spent working in rooms kept cold because of the instability, at higher temperatures, of the tissues he worked with.

Miescher's discovery of a particular type of chemical substance associated with the nucleus encouraged the idea that specific dyes, or stains, might be found to colour the material of the nucleus selectively, and make it easier to see under the microscope. This provided at least part of the impetus which led to the identification of the chromosomes, as bodies within the cell that take up colour. The first descriptions of chromosomes appeared during the 1870s, and by the 1880s mitosis had been observed and described. This was also about the time that Oskar Hertwig, in Berlin, and the Swiss Hermann Fol first observed the process of fertilisation in detail under the microscope, describing how the sperm cell penetrates the egg so that the nucleus of the sperm and the nucleus of the egg fuse together to create a new nucleus. And in 1881 Edward Zacharias showed that the chromosomes themselves were composed, at least partly, of Miescher's nuclein. As long ago as 1884, the zoologist Hertwig, a former student of Ernst Haeckel, was able to write: 'nuclein is the substance that is responsible not only for fertilisation but also for the transmission of hereditary characteristics'.* This statement strikes to the heart of the modern understanding of the role of DNA, and it was made only 14 years after Miescher's original paper was published. And yet,

* Quoted by Alfred Mirsky, 'The Discovery of DNA', in *Scientific American*, volume 218, page 78, June 1968.

within a further two decades nuclein had been relegated, in the opinion of virtually all cell biologists, to a peripheral role, seen merely as the scaffolding which propped up the all-important protein molecules. What went wrong?

It wasn't that there was any doubt about the reality of nucleic acid or its presence in the nucleus. Indeed, by the beginning of the twentieth century chemists had worked out the basic components of this nuclear material. The building block which was to give its name to the nucleic acids is ribose, a sugar built up from four carbon atoms forming a pentagonal ring with an oxygen atom, with another carbon atom linked to the ring in a side chain. The fact that a sugar was involved in the structure of the nucleic acids was known by 1900; just which sugar it was remained a matter of debate and a cause of confusion for three more decades, but that is no reason to conceal the evidence from you now. The pentose groups are held together by phosphate groups, each of which consists of a phosphorus atom surrounded by four oxygen atoms. Although many fanciful structures were suggested and found wanting before, much later, the correct structure was found, we now know that the phosphate group forms a link between the third carbon atom in one pentose ring (labelled in accordance with an arbitrary convention, counting round from the oxygen atom) to the fifth carbon atom in the next ribose molecule, the carbon atom in the side chain. The third type of building block which goes to make up a nucleic acid molecule is called a base.

Only five bases are found in nucleic acids, and all of them are based upon the familiar carbon ring structure, although, of course, there are many other chemical bases which have different structures. All of the five bases found in nucleic acids had been identified by the beginning of this century; they are called guanine, adenine, cytosine, thymine and uracil, and they are generally referred to simply by their initials, G, A, C, T and U. Later studies showed that one base is attached to each of the ribose groups in the chain formed by the linked ribose and phosphate groups – but that is getting ahead of the story. Two of the bases, guanine and adenine, are very similar to one another (see

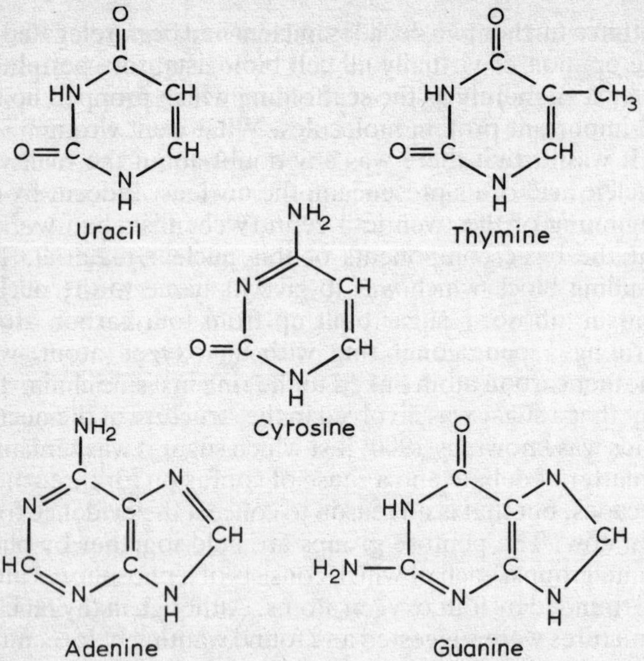

Figure 7.1 Five different bases are found attached to the sugar-phosphate chains of DNA and RNA. The order of these bases along the chain spells out the letters of the genetic code. C, A, and G are found in both DNA and RNA; T occurs only in DNA and U only in RNA.

Figure 7.1), with a double ring structure, and are both members of a family called the purines. The others, thymine, cytosine and uracil, have a somewhat simpler single ring structure, are also similar to one another, and are all members of the pyrimidine family.

The chemical groundwork for the modern understanding of the structure of DNA was laid by Phoebus Levene and his colleagues at the Rockefeller Institute for Medical Research in the first decade of the twentieth century. Levene himself must have been quite a character. He was born in Russia in 1869 (the year Miescher discovered nuclein), and although Jewish was allowed to join the Imperial Military Medical Academy in St. Petersburg,

becoming a Captain in the Russian Army before emigrating to the United States in 1891 as a result of increasing religious persecution. He never obtained a degree in chemistry, but came into the subject, like Miescher, through his interest in fundamental medical research, and he became one of the outstanding biochemists of his day. It was only in 1909 that this work led to the correct identification of ribose as the sugar in nucleic acid from yeast cells. At that time, and for years to come, it was widely accepted that the sugar in nuclei of cells from animals was a hexose, containing six carbon atoms, and so it seemed that there was a fundamental difference between plant and animal nucleic acids. It was only in the 1920s that the other form of nucleic acid was correctly identified as containing a very similar sugar to ribose, with one oxygen atom less, deoxyribose.*
And then it took some time for the realisation to sink in that both forms of nucleic acid occur in both animals and plants. Even then, nobody gave the nucleic acids so prominent a role in the life processes as Hertwig had in 1884.

## False dogma

There is one other difference between RNA and DNA, apart from the missing oxygen atom in the pentose. Each of them incorporates only four out of the five bases found in nucleic acids. DNA molecules contain G, A, C and T, whereas RNA molecules contain G, A, C, and U. In the 1920s and 1930s, biochemists knew that the nucleic acid present in chromosomes was DNA, but they also knew that

---

* The correct name for the nucleic acid containing ribose is 'ribose nucleic acid'. The correct name for the nucleic acid containing the pentose sugar which is produced when one oxygen atom is removed from ribose is 'deoxy ribose nucleic acid' (with a number inserted to tell you which oxygen atom is removed). Hardly surprisingly, these names have been contracted in common usage to ribonucleic acid and deoxyribonucleic acid, written in shorthand as RNA and DNA. Some pedants still use the longer versions of the names, which you may come across in other books. I will stick with the short versions from now on.

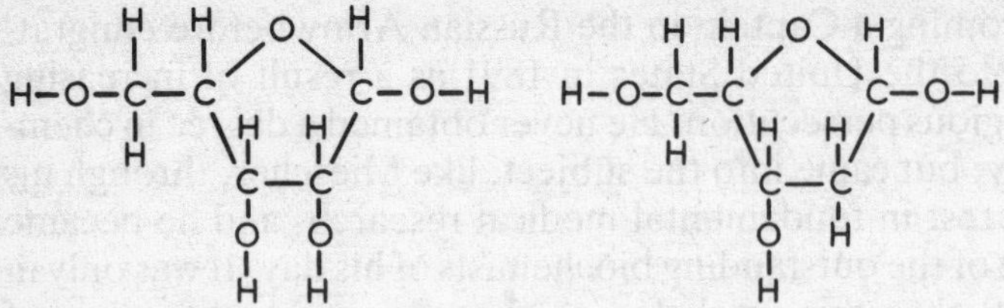

The five-carbon sugar in RNA      The five-carbon sugar in DNA
Ribose - $C_5H_{10}O_5$              Deoxyribose - $C_5H_{10}O_4$

Figure 7.2 Two almost identical sugars provide the backbones for the two fundamental molecules of life.

chromosomes contain protein in addition to the DNA. Protein looked much more interesting at that time, and it was generally assumed that the genetic material must be protein. The more it became clear that the cell, and the nucleus in particular, directed the production of enzymes for the body, the better this picture looked. After all, the enzymes were known to be proteins, and one way in which the cell could 'know' how to make proteins would be if it contained, in the nucleus, one or more examples of every protein it would ever need. Then, additional working copies of the protein molecules – enzymes and the rest – could be manufactured literally by copying the master copies. According to this image of the chromosome, the role of the nucleic acid – DNA – was merely that of holding the proteins together, providing the structure on which the protein molecules were hung.

In line with this idea, Levene developed the view that each of the four bases in each of the two nucleic acids is present in exactly equal proportions. This was the basis for models of DNA which saw the molecule as a chain built up of repeating subunits, pyrimidine followed by purine followed by pyrimidine, each attached to a ribose group, with the riboses held together by phosphate groups. One such unit (base plus sugar plus phosphate group) is known as a nucleotide; Levene's DNA structure became known as the tetranucleotide hypothesis, and although variations were suggested, including one which saw the four nucleotides

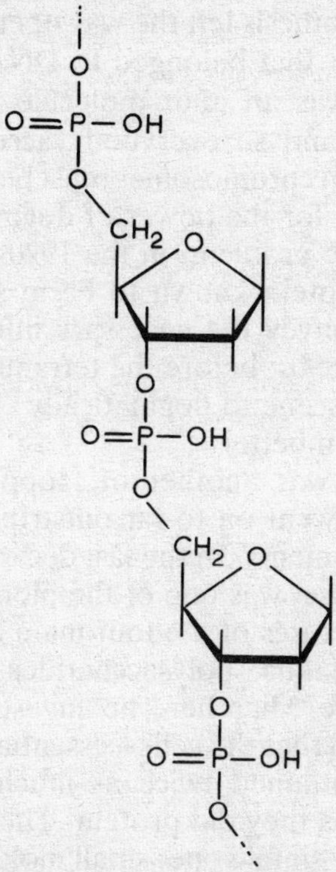

Figure 7.3 In both RNA and DNA, the sugars are joined to phosphate groups in an alternating chain.

linked in a ring, the essential point about the structure is that it is simple and repetitive. There is no scope, in the tetranucleotide hypothesis, for the DNA molecule to contain or convey information, any more than there is any meaning in the simple repetition of the letters GCAT GCAT GCAT GCAT . . . That's fine if all you are looking for is a structure to hold proteins in place on the chromosome. Unfortunately, Levene's hypothesis quickly gained the status of dogma, and prevented people from looking at alternatives until well into the 1940s. Levene was a brilliant chemist whose analyses of nucleic acids and investigation of their components helped to establish their separate identity, distinct from proteins. But his theoretical ideas were not so well founded, and misplaced faith in the tetranucleotide hypothesis left the way open for protein to be assigned the role that belonged to DNA. 'Everybody knew' that DNA was an idiot molecule made up of a repeating pattern, and so everybody accepted that the information stored in chromosomes must be in the protein.

It says something for the power of dogma that nobody seriously tackled the problem, in the 1920s and 1930s, of how the kind of protein known to be in sperm cells, in particular, could convey the necessary information. The researchers of the 1890s, before the tetranucleotide hypothesis came on the scene to dogmatically cloud the issue, could have told them better.

Albrecht Kossel was another of Hoppe-Seyler's students, although he went on to far outstrip his teacher in biochemical achievements. In the last decades of the nineteenth century, Kossel was one of the pioneers who analysed the basic structures of the four main components of living material, the fats, polysaccharides, proteins and nucleic acids.* Like Miescher, he investigated salmon sperm, and he noted that the cells – essentially just bags of chromosomes – contained twice as much nucleic acid, weight for weight, as they did protein. The protein in the cells is a particularly simple one, small molecules made up almost entirely of a single amino acid, arginine. This information, common knowledge even in the 1890s, ought to have been enough alone to show that the hereditary information in the sperm cells *must* by carried by the DNA. After all, if a message that runs repetitively GCAT GCAT looks stupid enough, a 'message' that runs *a a a* is positively moronic. Here in this simple protein we have *exactly* the kind of 'stiffening' needed to hold chromosomes together, while the bulk of the chromosomes, the all-important material in a sperm cell which must carry as little weight as possible, is the DNA.

If there was a need for further evidence, Kossel's work should have provided it. In ordinary salmon cells, the protein content of the chromosomes is another relatively

---

* He got the Nobel Prize for this work, in 1910.

simple form, called histone. Histone is simple compared
with most proteins in the body, but it is more complex than
protamine. If proteins were indeed made by copying master copies held in the chromosomes, how could the histones
have been produced, since the fertilised egg only had protamines to copy? And how could more complex enzymes
be manufactured by cells which only had histones to copy?
Although they were aware of the problem, biochemists in
the first 40 years of the present century either ignored it, in
the hope that it would go away, or trusted that later discoveries would reveal hidden complexity in the histones and
protamines. Even Kossel, whose career continued well
into the twentieth century, was seduced, by the obvious
complexity of most protein molecules and the apparent
simplicity of nucleic acid molecules (only four bases, compared with the 20 or so amino acids in proteins), into
making the same mistake, so we shouldn't be too harsh on
the lesser men and women who also failed to see the light.
After all, 'everybody knew' that the complexity they
needed to explain how a single fertilised egg could grow
into a mature, functioning adult simply couldn't be contained in those simple DNA molecules, with their repeating patterns of bases strung together along a sugar-
phosphate backbone.

# DNA at work

The central role of DNA began to emerge in 1928, as a
result of experiments carried out by the English bacteriologist Fred Griffith, using the organisms that cause pneumonia. Biologists who work with micro-organisms –
microbiologists – have a chance to see evolution at work
directly, because bacteria and viruses go through many
generations in a matter of hours. Changes that would
require years, or lifetimes, of investigation in plants such as
Mendel's peas, or even using *Drosophila*, can be observed
in a matter of weeks in a culture bred in the laboratory.
Griffith himself was not a geneticist, but a medical officer
working for the Ministry of Health in London. His interest

was in the pneumococci as agents of disease, rather than as a genetic research tool – but that didn't stop him noticing something of crucial importance to the development of genetics.

Griffith's work started from the description of different strains of bacteria as rough (R) and smooth (S) forms, by J. A. Arkwright, of the Lister Institute in London, in 1921. The key feature of Arkwright's observation was that the S form of several different bacteria is virulent, and causes disease, while the R form is weak, and produces little or no infection. The names mean just what they suggest – S forms of bacteria look shiny and smooth in cultures, thanks to a coating of a polysaccharide sugar in which the bacterium wraps itself. This coating seems to help it fool the body's defences, and enable it to produce infection. The R form lacks the smooth coating, so cultures of this bacterium look rough and lumpy, and it is readily identified as an invader and destroyed by the body it tries to attack. Following Arkwright's discovery of the phenomenon in a bacterium known as Shiga's bacillus, many bacteria were identified in the two forms, and Griffith reported the existence of R and S forms of pneumococci in 1923. Five years later, he reported something much more strange.

Griffith had been studying the virulence of different strains of pneumococci, using mice as his experimental subjects. The R form proved harmless when injected into the mice, while the S form was lethal. Seeking information which might lead to treatment for human sufferers from pneumonia, Griffith injected some mice with S pneumococci that had been killed (he hoped) by heat. Sure enough, the dead S bacteria were just as harmless to the mice as the live R bacteria were. But could the killed pneumococci become virulent again? Clearly, this was a question of great importance to the medical officer of health, and in the course of his further experiments Griffith tried injecting mice with a mixture of the dead, heat-treated S bacteria and a few live, but innocuous, R type pneumococci. The cocktail proved every bit as lethal to the mice as the original S strain had been, and when Griffith carried out a post-mortem examination he made his extra-ordinary discovery. What had killed the mice was a growth of live, smooth pneumococci. These virulent bacteria were teeming in the blood of the infected mice, and when they were transferred to a dish in the laboratory they continued to breed true, producing a viable colony of S type cells. The harmless R cells had not only been transformed into vir-ulent S cells by being mixed with the dead S pneumococci, but they had 'learnt' how to pass on their virulence to their offspring. Somehow, a gene belonging to the S pneu-mococci had become part of the hereditary material of the R strain. Griffith himself did not make this connection. But after his results were published in 1928 and confirmed by other researchers the genetically oriented microbiologists soon realised their importance.

One of those microbiologists was Oswald Avery, at the Rockefeller Institute in New York. Avery, who was born in 1877, had been working on pneumonia, which was then the main cause of death in people, full time since 1913. He and his colleagues were, at first, reluctant to accept Griffith's discovery, since it ran counter to a great deal of work at the Rockefeller which had established the exist-ence of distinct types of pneumococci; at first sight, Griffith's work seemed to show that there was less dif-ference between these strains than the Rockefeller work had indicated. But their own experiments, carried out by Martin Dawson, and those of other researchers, confirmed the accuracy of Griffith's work, and in 1931 the Rockefeller team showed that it wasn't even necessary to use mice as unwitting intermediaries in the transformation of one bac-terial form into another. Just growing a culture of R pneu-mococci in a glass dish (a Petrie dish) which also contained dead, heat-treated S pneumococci was enough to trans-form the growing colony from the R type to the S type, and provide it with the ability to kill mice. The transformation was real. But what was the transforming substance?

The next step on the trail to the identification of DNA with the genetic material was made by James Alloway, also in Avery's lab. He used alternate freezing and heating to break apart the cells of a dead colony of S bacteria, and then used a centrifuge to get rid of the cell debris, extracting the material from inside the cell. This neat trick, which was important in the later investigations of DNA, simply involves a machine which whirls a test tube of material round and round at high speed. The centrifugal force on the contents of the test tube causes the solid material to settle out at the bottom of the tube, and leaves light, liquid contents above the debris. The liquid extract alone was soon found to be sufficient to transform a growing colony of R cells into the S type. The transforming substance was part of the soluble fraction from inside the cell, not part of the solid debris. All of these early discoveries were being made in the early 1930s. By 1935 Avery, who had previously kept a professorial eye on the work going on in his lab, but whose name had not appeared on any of the research papers on the transforming substance published by Dawson and by Alloway, had decided that a concerted attack was needed to identify the transforming agent. He took on the task himself, with the aid of two junior researchers, Colin Macleod and Maclyn McCarty, and the three of them set about their task with such dedication and attention to detail that it took them nearly ten years to solve the puzzle. But, when their results were published in 1944, they left no room to doubt that the transforming agent – which must, almost by definition, be some of the genetic material of the bacterium – was DNA.

The team began by using a process of elimination, finding out what the transforming substance was *not* before trying to find out what it *was*. Perhaps it was a protein. So they attacked the active ingredients they obtained from dead S cells with enzymes that were known to chop proteins apart. The mixture still retained its transforming power, even though it no longer contained any workable proteins, so proteins could be ruled out – the first surprise. Perhaps the transforming power lay in the polysaccharides with which the S bacteria coated themselves. So the transforming agent was treated with an enzyme that broke up polysaccharides, again with no effect on its transforming power. So now Avery's team had to refine their product carefully, using delicate chemical techniques to remove all traces of protein and polysaccharides from their brew, since they were definitely not the active ingredients. They were left with pure transforming agent, isolated in a form in which it could be tackled head on by chemical analysis. It was not a protein or a polysaccharide, and the extraction method, involving alcohol, would have destroyed fats. There was only one thing left it could be, and that was nucleic acid. Sure enough, the chemical analysis revealed the tell-tale trace of phosphorus, and more subtle tests showed that the transforming agent was DNA, not RNA. Cautiously, the great 1944 paper did not specifically make the final step of identifying the DNA with the genetic material, though Avery did speculate along those lines in a letter to his brother Roy, a bacteriologist at Vanderbilt University.* Whatever the paper said for posterity, all of Avery's contemporaries were aware of the shocking implication, that DNA, not protein, carried hereditary information. The transforming agent was DNA at work. It was in 1944, at last, that nucleic acids took up their rightful position at the centre of the biological stage, where they have remained ever since. Avery was 67 at the time, a remarkably great age for someone to achieve such an important scientific breakthrough. He died in 1955, robbed of a certain Nobel Prize by his failure to live to an even riper old age. Griffith had been killed in the Blitz in 1941, in his sixty-first year, and never lived to see the culmination of the line of research his discoveries had initiated. But younger men were ready to pick up the torch.

---

* See, for example, Judson, page 39. Some people have speculated that Avery's work did not receive immediate, widespread acclaim, perhaps because of the legacy of Levene, a towering figure at the Rockefeller Institute, who had died only in 1940. After all, the one thing that these discoveries did above all else was to pull the rug from under the tetranucleotide hypothesis. But Robert Olby's authoritative history *The Path to the Double Helix* knocks this speculation on the head. There was isolated opposition to Avery's team's conclusions from a few individuals, but most biochemists, including those at the Rockefeller, were immediately impressed. When Avery first presented the key work formally to his colleagues at an after tea meeting in 1943, he received 'a standing ovation . . . a terrifically warm reception'. (McCarty, quoted by Olby, page 205.)

# Getting the right mix

Erwin Chargaff, born in Vienna in 1905, was the first major contributor to the DNA story to be born in the twentieth century. He moved to the United States in 1928, working at Yale University for a couple of years before returning to Europe, and then, in the mid-1930s, emigrated to the US, where he took up a post in the biochemistry department at Columbia University. During the Second World War, Chargaff was at the head of a small team investigating the biochemistry of the cell, especially the substances known as lipids and lipoproteins, which are combined fat and protein molecules. He was also involved with a study of the rickettsia virus, and as a result had some experience of working with nucleic acids, and kept up with the work on nucleic acids reported by other researchers. When Avery, Macleod and McCarty published their classic paper in 1944, biochemists in the United States very quickly became aware of the importance of their work.* And the person on whom the news seems to have made the deepest impression was Chargaff, who later told Robert Olby 'this was really the decisive influence . . . to devote our laboratory almost completely to the chemistry of nucleic acids.'†

Their first problem, however, was to get hold of enough DNA to work with. Even in the 1940s – and for some time afterward – it remained very difficult to extract DNA from cells and purify it, so that the chemists were faced with the task of analysing very small quantities of the pure material, and trying to determine the exact proportions of its constituents. That is why the tetranucleotide hypothesis had held

---

* This wasn't so outside this circle of specialists. The crucial results were never summarised for publication in a short paper in one of the weekly journals, *Science* or *Nature*, where key discoveries are usually disseminated to the wider world of science, and at the end of the war travel overseas was still difficult. The news did spread to Paris, where André Boivin confirmed the role of DNA as the transforming agent and there was great interest in the work. But in Britain, for example, the new discoveries made little immediate impact even on biochemists.
† *The Path to the Double Helix*, page 211.

out for so long. The techniques available simply were not accurate enough to determine precisely the proportions of the four bases present in a piece of DNA, and the available results were all in rough agreement with Levene's supposition that there were exactly equal numbers of all four bases present in all DNA molecules. At first, the task Chargaff had set himself looked hopeless. But just at this time Martin and Synge were developing the paper chromatography technique which they used with brilliant success to determine the amino acid sequences of proteins, and Chargaff realised that a variation of this trick could be used to analyse nucleic acids in a similar fashion.

The trick still wasn't easy. Chargaff's team had to find reagents that would chop off the pyrimidine and purine bases from the nucleic acids without significantly altering the chemical composition of the bases themselves, then they had to carry out the delicate chromatographic separations and analyse each component to find out the exact proportion of each base present in nucleic acid from the particular source. It soon became clear that the four bases were *not* present in exactly equal numbers. Different samples of DNA contain very different amounts of particular DNA bases, and there is just as much scope for a long molecule of DNA to contain information, written in the four letter GCAT code, as there is for protein molecules to contain information written in the 20 letter amino acid code. But it took a little longer for Chargaff to find out that there are simple rules which govern the ratio of one base to another, and over a period of several years the results appeared piece by piece in the scientific literature as each new experiment yielded a little more information.

In 1949, he could be found saying that 'a comparison of the . . . proportions reveals certain striking, but perhaps meaningless, regularities'; by 1951 he was a little less cautious, pointing out that 'as the number of examples of such regularity increases, the question will become pertinent whether it is merely accidental or whether it is an expression of certain structural principles'.* The regularities he

---

* Quotations reported by Portugal and Cohen, page 201.

was referring to were summed up in a paper published in 1950, and they can be stated very simply. The total amount of purine present in a sample of DNA (G + A) is always equal to the total amount of pyrimidine present (C + T), and the amount of A is the same as the amount of T, while the amount of G is the same as the amount of C. These are now known as the Chargaff Ratios, and they held the key to the next stage on the road to an understanding of the life molecule, the step which links Avery's work with the discovery of the double helix itself.

By 1952, the balance of evidence in favour of DNA being the life molecule – the molecule of heredity – was already impressive, although pockets of resistance to the idea remained. Most of those pockets were wiped out, and the scientific community at large made aware of the literally vital significance of DNA, by an experiment which lacked the painstaking accuracy and precision of the work of Avery and Chargaff, but had a brilliant simplicity and produced one clear, unambiguous result. The experiment was carried out by Alfred Hershey and Martha Chase, at the Cold Spring Harbor Laboratory on Long Island. It has gone down in history under the name of the most memorable piece of equipment used by Hershey and Chase, a household food mixer, as the Waring Blender Experiment.

Hershey and Chase were involved, like many other microbiologists, in the study of viruses which attack bacteria. These viruses, known as bacteriophages, or phage for short, are on the borderland between living and nonliving things, far smaller even than a bacterium. They infect bacteria and reproduce – replicate – inside the bacterial cells they have invaded, and phage experiments formed a natural progression in the world of genetic studies, where the experimenters had started out with large, long-lived organisms such as Mendel's peas and moved on to *Drosophila* and then bacteria. Each smaller type of organism reproduces faster and provides a chance to watch evolution at work on more generations in a shorter time; in addition, the smaller the organism the less

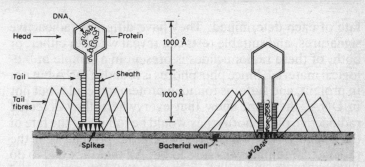

Figure 7.4 A phage. One Ångstrom is one hundred-millionth (10⁻⁸) of a centimetre.

extraneous material it carries and the more its genetic material dominates its structure, until viruses, including the bacteriophages, are scarcely anything more than a bag containing the genetic material.

Phages were pictured for the first time in the 1940s, using electron microscopy. A typical phage looks rather like a tadpole, with a head that is the bag of genetic material, and a tail which it uses to get about. Some have several 'tails', or spidery legs, which they use to attach themselves to a bacterium, punching a hole in the bacterial cell's wall and injecting their own virus genetic material into the much larger cell. A short time later, the bacterial cell bursts open and releases a flood of new phages into the world. What has happened is that the virus genetic material has hijacked the cell's machinery, turning it over to the manufacture of hundreds or thousands of new viruses, until all the chemical storehouses inside the cell have been used up. The bag of the virus 'head' is made of protein; the material inside – what we now know to be the genetic material – is DNA. But that wasn't known until the 1950s. It was the Waring Blender Experiment that convinced people that the virus achieves its ends by injecting DNA, not protein, into the bacterial cell, and that therefore DNA, not protein, must be the genetic material.

The experiment depended on growing phage in a medium which contained radioactive isotopes of phosphorus and sulphur. These radioisotopes can be traced through the cycle of infection and reproduction, and the fate of each determined. They have different radioactive signatures, and suitable tests can reveal whether either, or both, of these radionuclides is present in a sample of biological material. Since phosphorus is found in DNA but not in protein, and sulfur is found in protein molecules but not in DNA, the team knew that everywhere they detected radioactive phosphorus they would be following the fate of the phage DNA, while everywhere the protein went the radioactive sulphur was sure to be. All that they had to do was to take a culture of bacteria which had been attacked by the radioactively labelled phage and in which the genetic material had been injected into the bacterial cells, leaving the unimportant, genetically inactive material behind. By separating the infected cells from the phage leftovers on the skins of the cells, they could tell whether DNA or protein had been injected into the victims of the virus.

The snag was, they couldn't separate the phage leftovers on the skin of the bacterium from the active ingredient inside it – until, that is, a colleague lent them a domestic Waring Blender. The machine proved ideal, providing just enough agitation to shake the empty phage bags loose from the bacteria they had infected, without squashing everything, DNA and protein, into an amorphous mass. When the resulting mess was put into a centrifuge, the bacterial cells fell to the bottom and could be separated out, while the coats of the phages were left behind. Sure enough, as Hershey had speculated in advance, the radioactively labelled DNA was found in the infected bacterial cells, while the radioactively labelled protein was found in the cast-off coats. And one of the most important points, signifying the shift of opinion since 1944, is that that, indeed, was what Hershey *expected* to find. The theory, that phage operates by injecting DNA into a bacterium, came first, and the experiment was designed to test the theory. Avery *hadn't* expected to find that the transforming substance in pneumococci was DNA, but was led to that conclusion by the experiments.

In fact, the evidence from the Hershey–Chase experiment wasn't quite as conclusive as I have made it sound.

There was some contamination of the two components with one another even in spite of the separation, and there was room to doubt, if you wanted to doubt, that the genetic material had been *proved* to be DNA. But by 1952 the tide had turned, and most of the specialists in the field were either already convinced, or ready to be convinced, that DNA was the genetic material. So the Hershey–Chase experiment is important as marking the end of the transition from the idea that protein carries the genetic code to the idea that DNA carries the genetic code. Before Avery's work, scarcely anyone believed that genetic material was DNA; the balance shifted after 1944, tilting still further with Chargaff's work; after the Waring Blender experiment, scarcely anyone believed that the genetic material was *not* DNA. A substance that had been regarded as boring scaffolding eight years before now came under intense scrutiny as the fundamental molecule of life. Two questions were paramount: what was the structure of DNA? what was the nature of the genetic code? The answers began to come in very quickly, but not through traditional chemical techniques. The next step in the development of molecular biology owed a great deal to physics – both the abstract theoretical ideas of quantum physics, summarised by Erwin Schrödinger, and the entry of several trained physicists into research in molecular biology in the period shortly before, and especially just after, the Second World War.

## Schrödinger and the physicists

We have already met some of the physicists who played a key role in determining the structures of the molecules of life. Lawrence Bragg, who came to Cambridge as Cavendish Professor in 1937, and J. D. Bernal, who moved from Cambridge to London at the same time, became key figures in the study of proteins and other molecules using X-ray diffraction techniques, and in both places teams with the special skills for this kind of work quickly grew up. In Cambridge, Perutz (who originally came to Cambridge to work with Bernal during his time there) and Kendrew were just two of the physicists who were to make a major contribution to biology during Bragg's time at the Cavendish. In London, Bernal never achieved personally the success he had hoped for, either with protein structure or in determining the nature of viruses. His work was interrupted by the war almost before he could get started in his new base, and afterwards it was other researchers who picked up the threads, although Bernal remained a pioneering figure who had been the inspiration for many of the physicists working in biology in the 1940s and 1950s.

But the threads were picked up. At King's College, also in London, the post-war years brought the birth of a biophysics unit, the first of its kind, and for a long time the only one, in Britain. The unit was funded by the Medical Research Council, and headed by John Randall, who became Wheatstone Professor of Physics at King's in 1946. The college, located on the Strand in central London, was in a mess. The buildings had been used during the war as a base for fire engines, while the engineering laboratories had been turned over to munitions and the manufacture of machine tools. As if that wasn't enough, the Blitz itself had left its mark in the form of a crater 27 feet deep and 58 feet wide in the quadrangle. From these unpromising beginnings, Randall created a unit that was to play a major part in the search for the double helix, and one of the key members of that unit was Maurice Wilkins, a physicist who had worked on the Manhattan Project which developed the atomic bomb and had, as a result, become disillusioned with fundamental physics research. The MRC unit was running in 1947; in 1951, Wilkins was joined by an expert in the X-ray diffraction technique, Rosalind Franklin, of whom more later.

All of these people, and others working in Britain in the immediate post-war years, were essentially physicists investigating the structure of living things. At the same time, on the other side of the Atlantic, Pauling and his colleagues were doing much the same. Of course, all of these people were versed in the chemical studies, or carried out chemical analyses themselves. But the greatest success of the time, the alpha helix, came from the direct application of the laws of quantum physics and hydrogen bonding to determine the structure that best fitted the chemical evidence. Even Pauling, although generally labelled a chemist, made his greatest mark on science as a practising quantum physicist, whatever the label may have said.

But it wasn't just post-war disillusionment with the fruits of research in physics that drove physicists such as Wilkins into the arms of biology. There were very good, positive reasons why this kind of research should have been attractive to physicists at that time, and they had been spelled out in a little book by Erwin Schrödinger, one of the founding fathers of quantum physics, that was published in 1944. That book, called *What is Life?* made a deep impression on many physicists, and most of the main characters in the story of the discovery of the double helix have since recalled how it focused their attention on the possibilities of tackling biological problems from a physics standpoint. The structural approach to the molecules of life, involving X-ray diffraction studies and the Pauling method of making models, literally out of bits of paper, to see how molecules fitted together, represented, if you like, the engineering approach, the practical physics, in the sense 'practical' is used in school. Schrödinger's book provided a conceptual basis for that practical work, an underlying motivation and theoretical structure that gave the practical physicists something to aim at, and to dream about. It appeared at the perfect time to inspire a generation of researchers that may not all have been new to research, but who were mostly new to biology. But Schrödinger's little book had its own origins in the 1930s, and the work of Max Delbrück, one of the first physicists to turn to the biological puzzle of the nature of the molecules of life.

Delbrück was born in Berlin in 1906. Following up a boyhood interest in astronomy, he became a student of astrophysics, but switched to quantum physics in the late 1920s, just after the great breakthrough in quantum mechanics. In 1932 Delbrück was in Copenhagen when Niels Bohr, another of the quantum pioneers, gave a

famous talk in which he stated very clearly the view that the
nature of life could be explained on a purely physical basis,
and required no mysterious 'life force'. 'If we were able to
push the analysis of the mechanism of living organisms as
far as that of atomic phenomena,' he said, 'we should
scarcely expect to find any features differing from the
properties of inorganic matter.'*

The presentation by Bohr made a deep impression on
Delbrück. He soon took up a post in Berlin with Lise
Meitner, one of the first physicists to study radioactive
decay and a pioneering researcher in the study of nuclear
fission. The main reason for the move was, he said later,
the hope that the close proximity of the different Kaiser
Wilhelm Institutes in Berlin, each officially specialising in
one area of science but deliberately intended to encourage
cross-fertilisation of ideas, would enable him to become
acquainted with the problems of biology.† The hope was
soon fulfilled. Among the biologists Delbrück made contact with were N. W. Timofeeff-Ressovsky and K. G.
Zimmer. Together, the three of them produced a scientific
paper that was a milestone in the development of physical
ideas about heredity and the molecules of life.

The 'three-man paper', as it became known‡, concerned
the effect of X-rays in causing mutations in *Drosophila*.
Delbrück's contribution was a theoretical discussion of the
results which had been found by the other two authors, and
the key question the paper addressed was why a certain
amount of energy in the form of short-wave radiation
would cause a mutation in the genetic material, while the
same amount of energy in another form, such as the background heat of the lab in which the flies were raised, had no

---

* Bohr's address, 'Light and Life', was published in *Nature*, volume
131, page 421, in 1933.
† Quoted by Olby, page 231.
‡ Not to be confused with another famous paper, in quantum theory,
given the same German handle which translates as the 'three man
paper'. The shorthand name is the same, but the papers are different –
see my book *In Search of Schrödinger's Cat* for a mention of the other
one.

such effect. The puzzle struck at the heart of the basic dichotomy of genetics. On the one hand, genes had to be stable, copying themselves faithfully for many generations to ensure that like begets like. On the other hand, there must be scope for *occasional* changes, mutations, which provide the variability on which natural selection operates. And, once such a change has occurred, the new gene itself is then copied faithfully once again. How could this be?

Delbrück looked for an answer to the problem in terms of the known laws of quantum physics. He saw an analogy with the way a quantum of light – a photon – can make an electron jump from one state in an atom to another state, with nothing in between. Einstein's interpretation of the photoelectric effect provided that light must come in packets, as photons. A strong light of a certain frequency shone on to a metal surface produced more photoelectrons, but each of them had the same energy as the photoelectrons ejected when a weaker beam of light of the same frequency was shone on the surface. But each electron *does* get more energy when light with a shorter wavelength is used in the experiment. Each photon has only a certain amount of energy to give up, and the shorter the wavelength (higher the frequency) of the electromagnetic radiation the more energy each photon has to give. This provided the clue to mutations.

Genes, argued Delbrück, must be very stable molecules held together by strong forces between individual atoms. The sort of thermal energy involved in heat motion of the molecules at everyday temperatures wasn't enough to disrupt these strong bonds. But the energy carried by a highly energetic photon, an X-ray, could be enough to disrupt the genetic molecule and cause its structure to be re-arranged in a slightly different configuration. After the X-ray had done its work, the gene would be left in its new state, once again held together by the strong inter-atomic forces, to be copied faithfully by the cell and to pass on its mutated characteristics to the phenotype. Mutation is a quantum process involving molecules being pushed over an energy barrier from one stable configuration to a different stable configuration. And Delbrück and his colleagues went further. By using statistical techniques, they were able to get a rough idea of how big the sensitive region of genetic material must be in order to account for the observed effects of X-rays of a particular frequency on mutation rates in *Drosophila*. They came up with an estimate that the changes in genetic material which cause mutations involve less than one thousand atoms at a time.

In the middle of the 1930s (the three-man paper was published in 1935) these were significant new ideas. The *Drosophila* experiments, and their interpretation, provided some of the first direct evidence that genes were, in fact, molecules, and not more complex structures like sub-microscopic versions of the cell. (Remember that viruses were sub-microscopic in 1935, and they are much more complex than a gene.) But at that time nobody knew *which* molecules genes were made of.\* Avery was only just beginning to tackle seriously the puzzle of identifying Griffith's transforming substance in pneumococci, and even then he didn't work on the problem non-stop in the years that were to follow. Delbrück's key insights had to wait to find their true place, and it is a happy coincidence that Schrödinger's book was published in the same year that Avery's work came to fruition. But Schrödinger went further than Delbrück in setting physicists on the trail of the code of life.

Schrödinger knew Delbrück from his time in Berlin in the early 1930s, and the two were friends. By 1935, he had moved on, but Delbrück sent him a copy of the three-man paper. In the 1940s, Schrödinger, like so many of his

---

\* In fact, the physicists already had an important clue to the chemistry of the molecule of life. In a series of experiments that began in 1928, Lewis Stadler investigated the mutagenic effect of ultraviolet light on corn. He found that the wavelengths of light which are most effective at causing mutation are just the wavelengths that are absorbed by DNA, around 260 nanometers wavelength. This is significantly less than the region of wavelengths absorbed by protein molecules, which peaks around 280 nm. The implication is obvious with hindsight, but in the 1930s the scientific world wasn't primed to draw the obvious conclusion.

colleagues a refugee from Nazi Germany, had settled in Dublin where, in 1943, he gave a series of lectures that drew in part on Delbrück's ideas concerning mutation, and which were published as *What is Life?* the following year.* Part of the book deals specifically with Delbrück's model, and it was in this form, as Schrödinger's variation on a theme by Delbrück, that the idea reached a wide audience of physicists, many of them seeking pastures new, in the immediate post-war period. The book established the molecular basis of the units of heredity beyond reasonable doubt. But it did more than that.

Schrödinger introduced a concept 'that the most essential part of a living cell – the chromosome fibre – may suitably be called *an aperiodic crystal*'. (*What is Life?*, page 5.) He drew a distinction between an ordinary crystal of a substance such as common salt, where there is an endless repetition of a basic unit in a perfectly regular pattern, and the structure you might see in 'say, a Raphael tapestry, which shows no dull repetition, but an elaborate, coherent, meaningful design'. A periodic crystal, like one of common salt, can carry only a very limited amount of information, rather like, and for the same reason, Levene's tetranucleotide DNA. But an aperiodic crystal, in which there is structure obeying certain fundamental laws but no dull repetition, can carry an enormous amount of information. Schrödinger used the term 'code-script' for what we would now call the genetic code, or simply code. Of course, the idea was familiar to biochemists, not least from their work with proteins and the 20-letter amino acid alphabet. But Schrödinger introduced the idea to physicists, and in terms that any of them, familiar with basic crystallography since undergraduate days, could immediately relate to. In such structures, individual molecular groups, perhaps even individual atoms, were each every bit as important as the individual letters of the alphabet in which this book is written. As Schrödinger said (page 65) 'the number of [different] atoms in such a structure need

---

* The book is still in print and well worth reading.

not be very large to produce an almost unlimited number of possible arrangements', and he used the specific example of the Morse code, where 'the two different signs of dot and dash in well-ordered groups of not more than four allow of thirty different specifications'. With a third sign, in addition to dot and dash, used in groups of not more than ten, 'you could form 88,572 different "letters"; with five signs and groups up to 25, the number is 372,529,029,846,191,405'.

Schrödinger did not specifically address the question of whether the genetic code might be written in terms of the arrangement of amino acids in a polypeptide or the four 'letters' of the DNA bases. In 1943, when he was writing this book, the received wisdom was still that the active ingredient of chromosomes must be protein. But any physicist who picked up the idea in 1944 or later and was intrigued by the question of whether the DNA alphabet could be sufficient to store the information in the chromosomes could soon scribble down some rough numbers on the back of an envelope.

The sort of calculation many people made at the time might have gone something like this. In a polypeptide chain, there are 20 amino acid building blocks, a 20-letter alphabet. The number of different ways of arranging these letters is roughly $24 \times 10^{17}$ (24 followed by 17 zeroes), amply sufficient to carry the information necessary to account for the variety of proteins found in living organisms. DNA, on the other hand, has only four nucleotide bases, G, C, A and T, and at first sight a four-letter alphabet seems restrictive compared with the protein alphabet, even though it has a lot more scope than the two-letter Morse code. (And remember that we could, in principle, translate all of Shakespeare – indeed, the entire Library of Congress – into Morse code.) If we took 20 bases, five each of the four different varieties, we would have as many pieces to play with as before, but we would only be able to arrange them in $11 \times 10^8$ different ways, because of the inevitable repetition. In other words, a polypeptide chain can form more than two billion times as many different combinations of amino acids as a DNA chain *of the same length* can for different polynucleotide messages. But suppose the DNA chain is longer? Doubling the number of bases in the chain allows for just about as many different coded messages as in the 20-letter polypeptide chain. This is no restriction of any significance (a typical DNA molecule may, in fact, be five times as long as a typical protein molecule), and there is every reason to suppose that messages written in a four-letter code as bases strung out along molecules of DNA will suffice to carry all of the information required for the normal functioning of living organisms. The genetic code really is like Shakespeare on an Aldis lamp.

Schrödinger's book was influential for many reasons. It was timely, it was beautifully written in clear language, and the author himself was one of the most respected physicists of his day. Among the people who later specifically recalled the impact it made on their lives were Maurice Wilkins, who stressed the importance of the fact 'that Schrödinger wrote as a physicist',* Salvador Luria, who (together with Delbrück) founded a group devoted to studying the nature of heredity with the aid of phage experiments, Erwin Chargaff, who *did* calculate the number of possible sequences in a nucleotide chain, and stressed the link between Schrödinger's codescript and DNA at a symposium in 1950,† Hermann Staudinger, the Nobel Prize-winning German chemist who coined the term 'macromolecule',‡ and two men who were both new to biological research in the early 1950s, Francis Crick and James Watson.

# The searchers

Francis Crick was born in Northampton, England, in 1916. He graduated from University College, London, in 1938,

* Olby, page 247.
† Olby, page 214.
‡ Olby, pages 19–29

with a second class degree in physics, and started research intended to lead to a doctorate, but was interrupted by the outbreak of war. From 1940 to 1947 he worked for the Admiralty on the development of radar and magnetic mines; the experimental apparatus he had built during the early stages of his abortive doctoral work was destroyed by a German bomb during the war, but in any case by 1947 Crick's interests were moving in a different direction, towards the interface between biology and physics. He heard Pauling give a lecture in 1946, which fired his imagination, and, as we have already seen, he was also influenced towards studying biological problems by reading Schrödinger's book. As a preliminary stage on the road from physics to biology, he spent two years in Cambridge working on a study of the way magnetic particles move in cells, and this period gave him a chance to find out more about the kind of work being done in Cambridge. So it was in 1949, at the late age of 33, that Crick joined the Medical Research Council Unit at the Cavendish, headed by Perutz, and registered as a research student with the aim of getting a PhD for work involving X-ray studies of protein. He achieved the aim, submitting a thesis titled 'Polypeptides and Proteins: X-ray Studies' in 1953; but, shortly before the thesis was submitted, he and James Watson had solved the structure of DNA – six years after Crick moved from physics to biology, and less than four years after he joined the Perutz group at the Cavendish.

Crick was a late starter in biology, who more than made up for his late start by the brilliance of his work once he got going. Watson, whose name will always be linked with that of Crick in the history of science, could hardly have been more different. Born in 1928 in Chicago, he was by most standards a child prodigy, being allowed to enter the University there, on an experimental programme, to study zoology at the age of 15. He graduated in 1947, still only 19, and moved to Indiana University in Bloomington to begin PhD work on *Drosophila*. But he soon decided that the great days of *Drosophila* genetics were over, and shifted sideways to the study of phage, influenced by the presence at Indiana University of Salvador Luria, another European expatriate. Luria, born in Turin in 1912, had learned the techniques of phage research in Paris, and moved to the United States in 1940. Together with Delbrück, Luria founded a circle of researchers in different universities and research establishments around the US, who called themselves 'the phage group' and shared ideas in this important area of research. Summer schools held annually at Cold Spring Harbor enabled the members of the phage group to get together, and to pass on their ideas about the molecules of life to a wider audience of visitors, including many eminent physicists. Watson – who also read *What is Life?* at about this time – was strongly influenced by Luria, with whom he carried out his thesis work on 'The Biological Properties of X-ray Inactivated Bacteriophage'. By 1950, the 22-year-old Watson, a brash young man with a fresh PhD, was off to Copenhagen, where he worked for a time with Ole Maaløe on a project involving labelling phage DNA with radioactive phosphorus. The work was closely related to the way in which Hershey (also a senior member of the phage group) and Chase labelled phage, a little later, in the Waring Blender Experiment.*

But Watson had itchy feet. Having achieved so much so soon, he seems to have been restless to achieve even more with comparable speed (while Crick was equally eager to make up for lost time), and he was convinced of the central role of DNA as the life molecule. His age and background make this conviction understandable – while senior researchers such as Delbrück and Luria, or even Hershey, may have been *intellectually* persuaded by the evidence of the importance of DNA, they had grown up in science during the time when 'everybody knew' that genes were made of protein. Their hearts took more convincing than their heads. Watson only came to research after Avery's breakthrough had been achieved, and like all young people he was eager to embrace the new idea that overturned the

* Delbrück, Hershey and Luria shared the Nobel Prize for Physiology or Medicine in 1969, for their various contributions to the study of bacteriophages.

established wisdom of the previous generation. He just
*knew* DNA was the life molecule. In May 1951 Watson
attended a meeting in Naples, where he heard Maurice
Wilkins talking about X-ray studies of DNA. Convinced
that DNA was the genetic material and sure that the greatest discovery to be made in biology would be the determination of the structure of DNA, Watson, with no
knowledge of X-ray diffraction techniques, decided that
this was the way to solve the puzzle. He decided that he
should move to Cambridge, because he knew about Perutz
and his work on the structure of large molecules, and he
enlisted Luria's aid, by letter, to pull the necessary strings.
Luria's help was essential, both to persuade the Cavendish
group to find room for Watson and because Watson was
financed in Europe on a Merck Fellowship, only permitted
to carry out research in an area that met with the Fellowship Board's approval. The string-pulling was only partially successful. The board, dubious about this sudden
shift of direction by the 23-year-old prodigy, reduced the
grant for the second year of his fellowship from $3,000 to
$2,000, ending it in May 1952 instead of September.

Still, Watson was in Cambridge, where, thanks to a ploy
by Perutz, he was soon ensconced in comfortable rooms at
Clare College, registered as a research student. 'Working
for another PhD,' he said in *The Double Helix*, 'was nonsense, but only by using this dodge would I have the
possibility of college rooms.'* As for his accommodation
at the Cavendish, he was given space in the same room as
Francis Crick, the start of the collaboration that was to lead
them to the Nobel Prize. But they could never have
achieved this on their own. Above all else, they needed the
best X-ray diffraction data from DNA, and that meant the
data from Wilkins' group in London, whose pictures had
fired Watson's interest in Naples.

The first X-ray diffraction photographs of DNA had
been obtained by the pioneer William Astbury in 1938, but
there was then a gap of a dozen years before Wilkins' group

---

* *The Double Helix*, page 87.

tackled the subject again in the early 1950s. Wilkins was born in 1916, the same year as Francis Crick, but on the other side of the world, in New Zealand. He was brought up in England from the age of six, and graduated from the University of Cambridge in 1938. Unlike Crick, he did complete his PhD work, in 1940, before working during the war first on radar and then on the Manhattan Project. By 1950, he was Assistant Director of the Medical Research Council's fledgling Biophysics Research Unit at King's College in London, and it was in that year that he received a gift of some very pure DNA, prepared in the Bern laboratory of Rudolf Signer. This DNA was in the form of a gel, a kind of sticky goo, and when Wilkins poked it with a glass rod and pulled the rod away he noticed how 'a thin and almost invisible fibre of DNA was drawn out like a filament of spider's web'.* The perfection of these fibres suggested that the molecules in them must be arranged in a very orderly fashion, and Wilkins and a research student called Raymond Gosling quickly adapted their X-ray diffraction equipment (built from war surplus radiography machine parts) to take pictures of the patterns produced from the fibres. They obtained striking photographs, far superior to Astbury's. One key reason for this success was that they kept the fibres moist, whereas Astbury had worked with a dried film of DNA – an echo of Bernal's breakthrough with protein crystals kept in their mother liquor.

The beautiful pictures obtained from the fibres of DNA, compared with the different pattern of spots obtained from dried film, suggested to Wilkins that he was seeing a trans-formation similar to the change from the $\alpha$ to the $\beta$ forms of keratin described by Astbury. The problem now was to interpret the pattern of spots they obtained in terms of a structure for the DNA molecules. Although the basic pattern, with prominent rows of spots forming the arms of a cross, was very suggestive of a helical arrangement the

---

* Quotation from Wilkins' Nobel Address; see Portugal and Cohen, page 238

evidence was not yet compelling, and the researchers were still a long way from knowing just what sort of a helix it might be – might it be a single strand, or two or three strands twined together? If it was a helix, what was the pitch of the 'spiral staircase', and what held the molecule into that shape? and so on. At this early stage, Wilkins leaned towards the idea of a single helix. But, clearly, the team at King's needed more expertise in interpreting X-ray diffraction patterns. It came in the form of Rosalind Franklin.

Franklin, born in 1920, took a degree in physical chemistry in Cambridge and then worked on the structure of coal and compounds derived from coal. From 1947, she had been working in the Laboratoire Centrale des Services Chimiques de l'État in Paris; her work there helped to establish the basis of what is now carbon fibre technology. She was an expert in X-ray diffraction techniques, but knew no more than the average physical chemist about biological molecules; although she enjoyed her work in Paris, she felt that it was time to get back to England (partly for family reasons), and was put in touch with Randall by Charles Coulson, Professor of Theoretical Chemistry at King's College. Her skills were just what the team needed, and she joined the group at the beginning of 1951, on the understanding that she would take over the DNA studies, working with Gosling. Perhaps this was a misunderstanding. Franklin herself is now dead, and the other participants in the story naturally have their own opinions, but it does seem that Franklin initially believed that DNA would be 'her' project, and that Wilkins had no further interest in it. When it emerged that Wilkins did indeed have an interest, and expected Franklin to be not even an equal investigator but something of a junior partner, a personal animosity developed which did nothing to help the King's team solve the problem, and may well have left the way open for Watson and Crick. Reluctantly, under pressure from Randall, Wilkins agreed to leave the Signer DNA to Franklin and to concentrate himself on a sample provided by Chargaff. Undoubtedly his animosity towards Franklin was not improved when he found that this sample did not give such good diffraction patterns, and was much more difficult to work with.

Nor was everything smooth sailing in Cambridge. There, the friction was between Crick and Bragg – potentially a very serious situation for a research student in his mid-thirties, yet to make a mark in science, confronting the head of the Cavendish Laboratory. The story of how Crick was on at least two occasions forbidden by Bragg to work on DNA,* of how Wilkins found a release from his conflict with Franklin by talking about DNA with his old friend Crick and with Watson, and all the other tribulations of the search for the double helix in the early 1950s has been told in graphic detail by Watson in his book *The Double Helix*, a one-sided account, particularly unfair to Franklin, but an invaluable memoir. The story from what might have been Franklin's point of view – she died of cancer in 1958 – has been told by Anne Sayre in *Rosalind Franklin and DNA*, a book which ought to be read by everyone who knows Watson's account. Judson's *The Eighth Day of Creation* is more complete, and almost as entertaining as Watson, while Olby's *The Path to the Double Helix* provides the greatest weight of factual information. I shall pick out a few highlights here.

The background to all the activity in Cambridge and at King's in London was one of widespread interest in bio-molecules. Kendrew and Perutz were at a peak of activity in their work on proteins, and the appearance of the eight papers on the alpha helix by Pauling and Corey in April

---

* There were good reasons for this rather odd-looking instruction. In the post-war years, funds for fundamental research in Britain were extremely limited, and both Perutz's group at the Cavendish and Randall's biophysics unit at King's were funded by the same organisation, the Medical Research Council. There was every reason to avoid a duplication of effort which might represent a waste of limited resources, and it wasn't as if there weren't enough interesting problems in molecular biology to go round. This was the origin of the gentlemen's agreement between the Cavendish and King's College, an understanding which explicitly gave the King's team first crack at DNA. The problem was that Watson, in particular, and Crick, to a lesser extent, were not gentlemen.

1951 excited everybody investigating the structure of bio-molecules, and ensured that helices were tested as possible models for any tricky molecule then ·being investigated.* Pauling himself made a desultory (by his standards) attempt at explaining DNA in terms of a triple helix, apparently causing mental anguish to Watson, who had a great respect for Pauling's ability and feared that he and Crick were about to be beaten to the post in what he saw as a race – though, in fact, nobody else seems to have thought there was a race on, and Pauling was handicapped by having only Astbury's old photographs to work with.

Pauling's influence on Watson, in particular, went far beyond the idea of the alpha helix. Watson tells how Pauling's book *The Nature of the Chemical Bond* was indispensable to him and Crick, their prime source of information about the sizes and shapes of the molecules that form the sub-units of DNA. But equally important was Watson's faith in the Pauling approach of model building. Whereas Franklin's approach was analytical, measuring angles and intensities off the diffraction patterns and trying to interpret them in terms of bond lengths and so on by means of detailed mathematical analysis, Watson's idea was to try to fit the pieces together, like a jigsaw puzzle, to see how the molecules might be held together in helices, and then to 'predict' what the diffraction pattern ought to be, fine-tuning the model until it fitted the observed pattern.

Watson, then, brought to the joint work with Crick a characteristic approach to this kind of problem, an approach directly inherited from the Pauling school. But he didn't bring with him much knowledge of the kind of biochemistry needed for an understanding of DNA. Crick knew all the mathematical tricks of interpreting the diffraction patterns – the key trick is called Fourier transformation – but he, too, lacked the kind of detailed familiarity

---

* On the other hand, the announcement of the results of the Hershey–Chase experiment made no great impact on the Cambridge team. Watson needed no convincing that DNA was the molecule of life, and, indeed, he was already familiar with this kind of phage experiment. At best, to him, the Waring Blender results were no more than the icing on Avery's cake.

with chemistry that enabled Pauling to see quickly to the heart of a problem of this kind. The great thing they had going for them was that, as a team, they could bounce ideas off each other, pulling each other's notions to bits and rebuilding them as necessary. And that, of course, is what Wilkins and Franklin, working in the same building as each other in London but scarcely on speaking terms, so conspicuously lacked.

# On the trail

When Franklin arrived at King's early in 1951, the first thing she had to do was set up a proper X-ray diffraction lab. The jerrybuilt collection of spare parts that had been used by Wilkins and Gosling the year before had long since burned out, and it took eight months before she had the new equipment set up and working. By November, however, she had begun, with Gosling, to obtain very good results from the experiment, sufficient to give a talk at King's – a colloquium – on the progress so far. The notes for her talk, which have been preserved with her other papers, clearly indicate that at this meeting she specifically drew attention to the indications in the X-ray diffraction pattern that DNA molecules are helical, and that the probable structure of the helix was with the phosphate-sugar backbone outside and the nucleotide bases tucked away on the inside. Watson was present at this talk, but failed to take notes and, judging by *The Double Helix*, remembered the meeting chiefly for the fact that the King's people seemed unenthusiastic about model building. He went back to Cambridge with, apparently, no conscious recollection of the important indications that the DNA structure was a helix with the backbone outside, and he was able to report what he had heard only in the vaguest and incomplete terms to Crick. On the basis of these misunderstandings, Crick and Watson went off half-cock, building a three-stranded helical model of DNA with the phosphate groups in the centre of the structure. Eagerly, the duo reported their 'success' to Wilkins, who brought Franklin,

Gosling and two other colleagues up to Cambridge from London to see the model. But the model was hopelessly inconsistent with the X-ray data, and among other things Watson had made an error of a factor of ten in the amount of water in the structure. Franklin and Gosling, in particular, were not amused, and left Watson and Crick in no doubt about what they thought of the model. For once somewhat abashed, the pair retreated from their obsession with DNA for most of the early part of 1952, Crick doing his 'proper' work on proteins while Watson began a study of the tobacco mosaic virus.

But that didn't stop them speculating about the nature of DNA, and the central highspot in the Watson–Crick saga can be seen, with hindsight, as an idea that emerged almost casually from a conversation between Crick and John Griffith (the nephew of Frederick Griffith) who was a mathematician working in Cambridge at that time. The conversation took place in June 1952, after the two of them had been to a talk at the Cavendish by the astronomer Thomas Gold. Gold had been propounding the Steady State theory of the Universe, the idea that the Universe not only presents the same general aspects to an observer any*where* in it, but that it also presents the same general aspect at all *times*. This idea is now out of fashion, but it was a strong contender in cosmological debates in the 1950s, and the idea of a Universe constant in both time and space was, rather extravagantly, dubbed the Perfect Cosmological Principle. Crick and Griffith, idly discussing whether there ought to be a Perfect Biological Principle, decided that the appropriate candidate would be the mechanism by which the gene is able to copy itself – self-replication.

Crick tossed out the idea that the replication might involve the flat nucleotide bases in the DNA molecule. His idea was that the flat molecular groups might stack one on top of another, interleaving like two decks of cards riffled together to hold the different chains of DNA together, and he suggested that Griffith might care to work out which of the bases might attract one another and so form a stable structure. A little later, in an equally casual conversation, Griffith mentioned that he had, indeed, found that the bases paired up in specific ways. Adenine attracts thymine, and guanine attracts cytosine. Crick immediately appreciated the possible importance of this for replication. If you have a pair of molecules CT and another pair AG, and pull each of them apart, the loose C will pair up with another T, the loose T with another C, and so on, so that you end up with two sets of CT and two sets of AG. Repeating the process down the length of a DNA chain you can see at a glance how a molecule of DNA might replicate.

But with the prize they sought almost in their grasp Crick and Watson fumbled their first chance for glory. Crick did not appreciate then that Griffith's calculations involved the purines and pyrimidines oriented edge on to one another, not stacked one above the other, nor did he realise that the attraction between the complementary bases involved hydrogen bonds. Griffith had actually already been working along these lines on his own, but he was very much a junior member of the Cambridge fraternity, a mathematics graduate who was taking the undergraduate course in biochemistry as a shift in direction. He was too reticent to push forward his own biological ideas, and these, very much the precursor to the present understanding of the double helix, were never published. Nor, at the time, did Crick appreciate that the pairing of A with G and C with T immediately explained Chargaff's rules. Although Watson claims to have mentioned the rules to Crick, he must have forgotten them. It came as a bolt from the blue when Chargaff, on a visit to the Cavendish in July 1952, was introduced to Crick by Kendrew and explained the 1:1 ratios:

I said: 'What is that?' So he said: 'Well it is all published!' Of course I had never read the literature, so I would not know. Then he told me, and the effect was electric. That is why I remember it. I suddenly thought: 'Why, my God, if you have complementary pairing, you are bound to get a one to one ratio.' By this stage I had forgotten what Griffith had told me. I did not remember the names of the bases. Then I went to see Griffith and I asked him which his bases were and wrote them down. Then I had forgotten what Chargaff had told me, so I had to go back and look at the

literature. And to my astonishment the pairs that Griffith said
were the pairs that Chargaff said.*

In this way are some Nobel Prizes, at least, won.

Yet, with all of this evidence to hand, Watson and Crick
made no immediate attempt to build the definitive model
of DNA. In those days, says Crick, 'We could afford to
have a beautiful idea like that and just let it sit around for a
year. And just talk to our friends, and nobody else would
know.'† They could afford this luxury in no small measure
because Franklin had, around this time, discarded the idea
of helical DNA and was leading herself up a blind alley.
She had achieved an enormous amount on the practical
side, obtaining still better X-ray diffraction pictures and
distinguishing clearly between the two forms of DNA,
labelled A and B,‡ but she probably needed a breathing
space in which to get her theoretical ideas straight, and she
seems to have been led off the trail, temporarily, by trying
to determine the structure initially for the form of DNA
that did not show the clearest helical pattern. Whatever the
reasons, it was early 1953 before the DNA story suddenly
came to the boil. Then, everything happened in a rush.

# The double helix

In December 1952, it seemed that this luxurious delay had
proved fatal to the ambitions of the Cambridge team. Peter

---

* The whole Griffith episode is told in all the standard histories. Olby's
is the most complete, and this quotation from Crick appears on page 388
of *The Path to the Double Helix*; it is part of a longer interview recorded
by Olby in 1968.
† Crick, reported by Judson, page 144.
‡ This may have been partly because she now had even better material
to work with. Ralph Barclay, in Colorado, recently pointed out that he
had prepared DNA which was sent to Wilkins and passed on to
Franklin, and that he had used 'an isolation procedure that resulted in a
product, clean and pure enough to give Franklin the beautiful, well-
defined X-ray diffraction pattern . . . without that DNA, I had been told,
such pictures would not have been possible at that time.' (*New Scientist*,
8 March 1984, page 47.)

Pauling, the son of Linus, was a graduate student at the Cavendish and a friend cultivated by Watson. He received a letter from his father mentioning that he (Linus Pauling) and Corey had worked out a structure for DNA. In January 1953, Peter received a copy of the paper, which was scheduled to appear in the February issue of the *Proceedings of the National Academy of Sciences*. With sinking hearts, Watson and Crick saw that the model was a triple helix, with the phosphate groups – the molecular backbones – on the inside. At first, it looked as if they had been on the right trail after all. Then, to their astonishment, they realised that Pauling and Corey had made a blunder as significant as Watson's blunder, the one that had drawn such scorn from Franklin and Gosling when they had proudly presented their own three-chain model to the King's group. Amazed at this untypical error by the great man, Crick and Watson agreed that they must have one last fling, attempting to come up with the right structure before Pauling had the mistake pointed out to him and returned to the fray. They reckoned they had no more than six weeks to do the job, and the resulting bout of feverish activity is at the heart of Watson's presentation of the story of the double helix as a race. But the people who were really pipped at the post in the next few weeks were Franklin and Gosling. Ironically, if Peter Pauling had not been in Cambridge, and if Watson and Crick had not seen the PNAS paper until copies of the journal arrived in England in March 1953, then Franklin's correct interpretation of the structure of DNA as a double helix might already have been in the press with a journal such as *Nature* before the Cambridge group could have been galvanised into action.

A few days after receiving the copy of Pauling's paper, Watson took it down to London to show to Wilkins and Franklin. According to Watson (*The Double Helix*, page 96) his presence in her lab and the mention of helices roused such anger in Franklin that he retreated to Wilkins' protection. In the conversation that followed, Wilkins told Watson of the best DNA pictures yet obtained, the so-called B, or wet form. Furthermore, he gave Watson a print of one of Franklin's photographs – a distinct breach of etiquette, especially considering how strained Wilkins' own relations with Franklin were. 'The instant I saw the picture,' says Watson (page 98), 'my mouth fell open and my pulse began to race. The pattern was unbelievably simpler than those obtained previously . . . the black cross of reflections which dominated the picture could arise only from a helical structure.' This is not a point that could possibly have been lost on an experienced crystallographer such as Franklin. The B form was obviously a helix, and the pictures from it were relatively simple. The A form, on the other hand, produced more complicated photographs, much harder to interpret but containing, potentially, a lot more information. It was natural that she should have concentrated on the A form, and equally natural that Watson, the model builder seeking a quick solution, should have (proverbially) leapt upon the 'new' pictures from the B form with cries of delight. Little did he know that the picture Wilkins gave him had actually been obtained by Franklin back in May 1952. In addition, Watson took back with him to Cambridge the news that the X-ray data and density measurements fitted the possibility that DNA could be a double chain – just the structure required for complementary self-replication. It was then Friday, 30 January 1953.

Back in Cambridge, the model building started in earnest the following week. Using scale models of the different components of DNA, produced by the Cavendish machine shop,* Watson built the structures while Crick pointed out their deficiencies. With almost obstinate blindness to the significance of the Chargaff rules and Griffith's calculations, at first Watson tried yet again to produce a structure, this time a *double* helix, with the backbones of the molecules inside and the bases sticking out, and it was

---

* For this, of course, they needed Bragg's approval to work officially on DNA. Pauling's paper provided the key to getting this approval – competition with King's was officially out of the question, but Bragg would be happy to see the Cavendish beat Pauling to the structure. The repercussions of Pauling's triumph in an earlier 'race' with Bragg still echoed down the years.

only when that failed to fit the new X-ray data* that he tried, almost reluctantly, turning the model inside out so that the spines ran around the outside and the bases projected into the middle. Now, the bases had to be made to fit together very precisely, a genuine three-dimensional jigsaw puzzle. Still almost wilfully avoiding the right path, they tried to match up like with like (A to A, C to C, and so on). It was now 20 February, and three of the six weeks' leeway that Watson had imagined the Cavendish team to hold over Pauling had gone. Ironically, they were saved from further delay by a visitor to Cambridge, the American Jerry Donohue, a former colleague of Pauling and an expert on hydrogen bonding.

The hydrogen atoms on the outside of organic molecules, such as the nucleotide bases, cannot always be assigned to a specific location with confidence. Just as a bond may be single or double, or have a flavour of both, being in effect a 1.5 bond, so the hydrogen nuclei – protons – may shift position, perhaps 'belonging' to an oxygen atom on the fringe of a benzene ring, or perhaps preferring to be located next to a nitrogen atom, for example. We saw something similar in Chapter Five with resonance and hybrid bonds. Just where the outer protons ought to be placed in drawing out the geometry of a molecule or group such as thymine or guanine (or making cardboard models) depends on the detailed quantum mechanical interpretation of the structure. Although the standard textbooks had led Watson and Crick to assume one particular form for these two bases, the so-called keto tautomers, Donohue knew that they were in fact almost certain to be found in the alternative, enol, form, a form which is much more suited to making hydrogen bonds. It took a week for the message to sink in properly. Then, playing with his cardboard cutouts representing the four nucleotide bases,

---

* The data from King's, of course, told Watson and Crick far more than the basic nature of the DNA molecule as a double helix. The diameter of the molecule, the angle of the 'thread' of the helical screw and other parameters were all constrained by Franklin's observations.

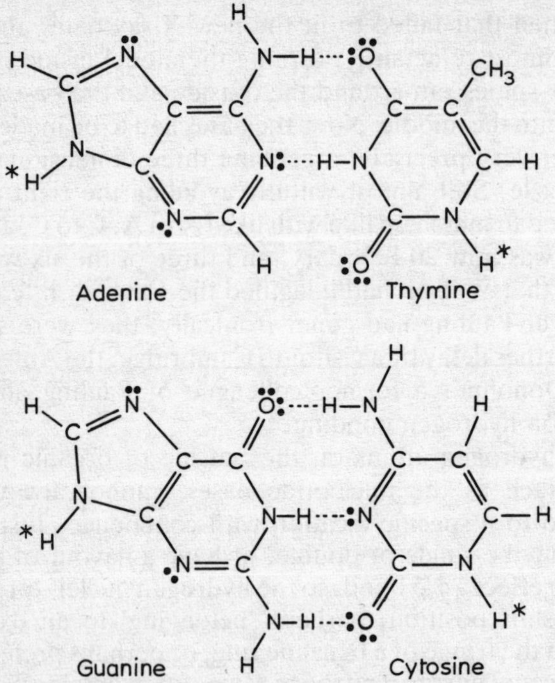

Figure 7.5 Hydrogen bonding between A−T and between G−C operates with the precision of two- or three-pin plugs matching up with their appropriate sockets.

Watson finally twigged that an adenine-thymine pair held together by two hydrogen bonds was identical in shape to a guanine-cytosine pair held together by two or more hydrogen bonds (Pauling and Corey later showed that there are in fact three hydrogen bonds between these two bases). Donohue confirmed that these structures made sense:

> My morale skyrocketed, for I suspected that we now had the answer . . . the hydrogen-bonding requirement meant that adenine would always pair with thymine, while guanine could pair only with cytosine. Chargaff's rules then suddenly stood out as a consequence of a double-helical structure for DNA. Even more exciting, this type of double helix suggested a replication scheme much more satisfactory than my briefly considered like-with-like pairing. Always pairing adenine with thymine and guanine with cytosine meant that the base sequences of the two intertwined

chains were complementary to each other. Given the base
sequence of one chain, that of its partner was automatically determined. Conceptually, it was thus very easy to visualise how a
single chain could be the template for the synthesis of a chain with
a complementary sequence.*

So Watson and Crick came up with their famous
model of DNA as a double helix in which the complementary nature of the two strands provides the mechanism by
which the gene can copy itself. During the first week of
March, an accurate model was built using the metal plates
from the Cavendish machine shop, shaped to represent the
quantum mechanically correct versions of the molecules
and arranged at the correct positions and angles to fit the
X-ray data. The helicity and the double-stranded nature of
the molecule were already implicit in those X-ray data; the
key contribution that came from the Cavendish team was
the idea of the base pairing which held the strands
together. The exact agreement between the resulting structure and the X-ray diffraction evidence was vital in proving
that brilliant insight correct. The X-ray data, of course,
themselves depended on the laws of quantum physics, and
could only be interpreted with the aid of a quantum view of
the world of atoms and molecules. But the brilliant insight
which Watson and Crick achieved (with some help from
Donohue, Griffith and Chargaff) was more fundamentally
rooted in quantum physics. It depended upon the actual
shapes of the bases attached to the sugar-phosphate backbone of a DNA molecule, and those shapes could only be
understood in quantum mechanical terms, following the
guidelines originally established by Pauling. And even
when the correct shapes were built in to the models, the
links between A and T and between C and G turned out to

---

* Watson, *The Double Helix*, pages 114 and 115. The statement that
'Chargaff's rules then suddenly stood out', more than half a year after
Crick's conversations with Griffith and with Chargaff himself, is
astonishing, and perhaps gives the lie to the idea that Watson and Crick
had already come up with the complementary idea and were hugging it
to themselves in delight rather than sharing it with the world.

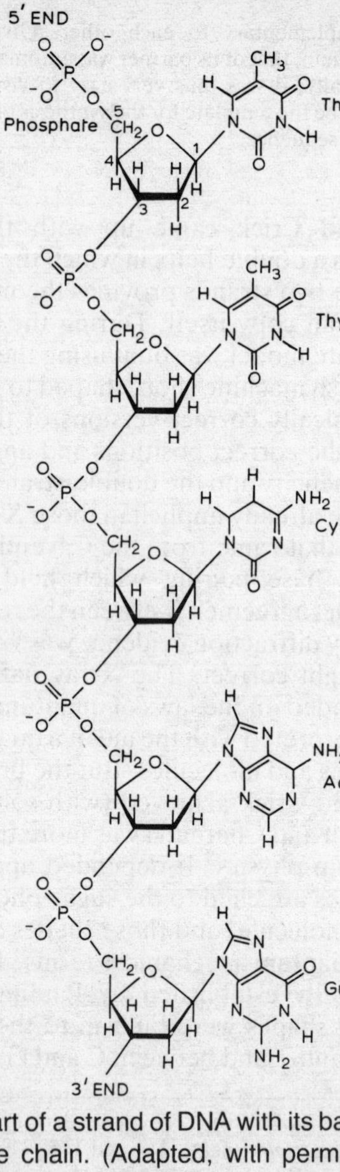

Figure 7.6 Part of a strand of DNA with its bases attached to the sugars of the chain. (Adapted, with permission, from Figure 14.10 of *Biology*, by Helena Curtis, published by Worth, New York, 1975.)

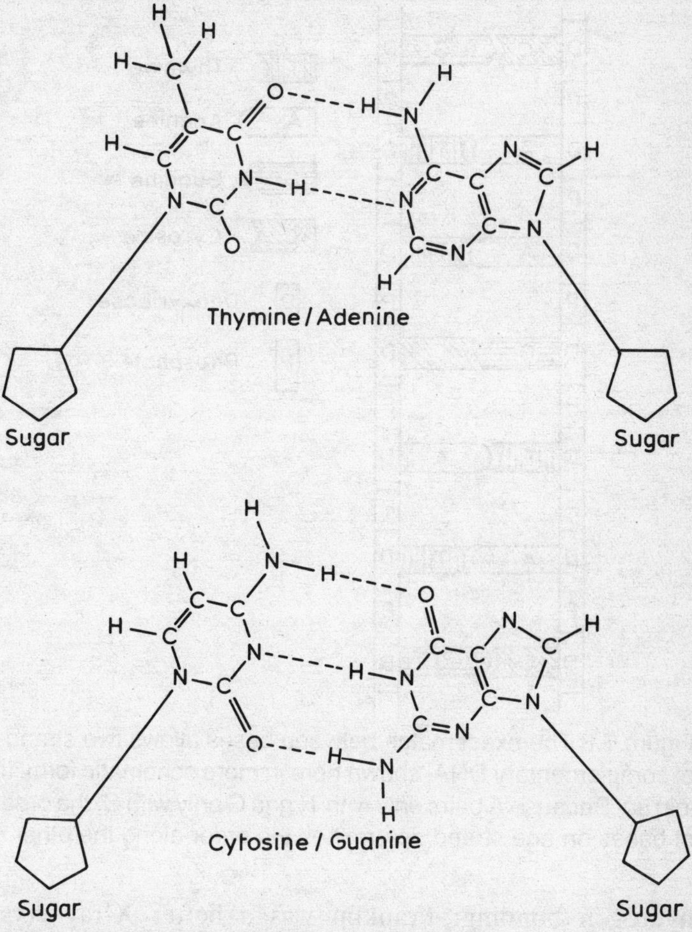

Figure 7.7 When the paired bases on complementary strands of DNA join up through hydrogen bonding, the T–A bridge is exactly the same size and shape as the C–G bridge.

be hydrogen bonds, themselves a feature of the quantum behaviour of matter. Quantum physics lies at the heart of life.

As even this brief outline of the story shows, Crick and Watson were not real experts in any of the areas of science that came together to provide a picture of the double helix. Donohue knew more about the shapes of molecules and

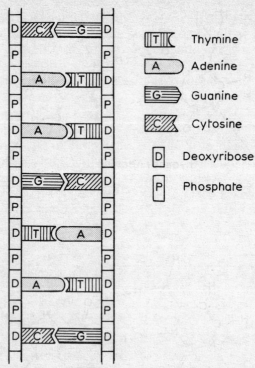

Figure 7.8 The exact match between bases allows two strands of complementary DNA, shown here in more schematic form, to join up. Because A pairs only with T, and G only with C, the order of bases on one strand determines the order along the other.

hydrogen bonding; Franklin was a better X-ray crystallographer; Chargaff understood the relationship between the bases; and so on. But what Watson, in particular, contributed was an ability to see the broad picture, to take what was needed from several specialist disciplines and come up with something new that was greater than the sum of its parts, and which none of the specialists, unable to see the wood for the trees, was quite able to perceive.

The contribution made by Crick and Watson was original, important and enough to have enshrined their names forever in the annals of scientific fame. It is quite clear today, looking back, how much they owed to the King's data, and it is clear now – as, surely, it must have been then – just how they should have set about publishing their conclusions. Scientific etiquette suggests that in circumstances like this, where a team of theorists has used an experimenter's unpublished data to come up with a new insight, the right thing to do is arrange some sort of joint publication. In this case, the obvious thing to do, other things being equal, would have been for Crick and Watson to get in touch with Wilkins and Franklin, giving them the good news, and for a joint paper with all four names on to be prepared, telling the world that, first, the X-ray data showed DNA to be a double helix and, secondly, that a new theory of base pairing could explain how the helix was held together and how it replicated. In such a paper, Gosling, the humble graduate student, might have received a mention in the list of credits. The obvious alternative would have been for two papers to appear alongside each other in the same journal – one from Wilkins and Franklin giving the X-ray data, followed by one from Watson and Crick giving their interpretation of it. But the eagerness of the Cambridge team, and the antipathy between Wilkins and Franklin, combined to produce instead a distinctly odd-looking series of publications in the journal *Nature*.

## Franklin's near miss

The full model of DNA was complete on Saturday, 7 March 1953. The next Monday, Crick received a letter from Wilkins informing him that Franklin was leaving King's to work with Bernal's group at Birkbeck College, and that since 'much of the 3 dimensional data is already in our hands'* he expected to solve the structure of DNA very shortly. The 'race' had indeed been close; even Wilkins, however, did not immediately appreciate how close. During that second week in March, news of the great

* Letter dated 7 March; quoted by, for example, Olby, page 414.

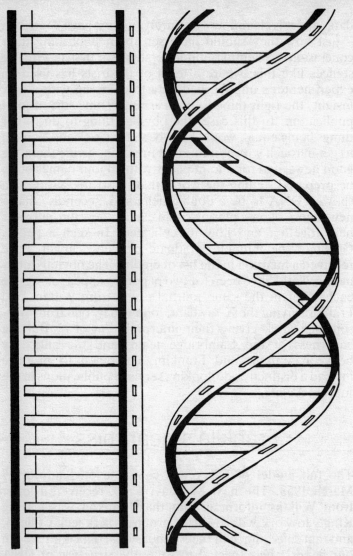

Figure 7.9 In fact, the two DNA strands are twisted around each other to make a double helix.

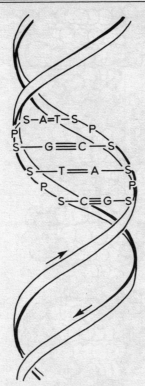

Figure 7.10 A closeup of part of a DNA double helix.

discovery spread, by word of mouth around Cambridge and by letter to researchers such as Delbrück and Pauling; by the weekend, the draft of a short paper to *Nature* had been prepared, and a copy was sent off to King's. Wilkins was magnanimous. 'I think you are a couple of old rogues,' he wrote back on 18 March, as soon as he had seen the draft of the *Nature* paper, 'but there is no good grousing – I think it is a very exciting notion and who the hell got it isn't what matters.'* He went on to suggest that he (Wilkins) and his colleagues might publish a short note alongside Watson and Crick's in *Nature*, 'showing the general helical case', and right at the end of his letter mentioned that Franklin

* See Olby, pages 417 and 418.

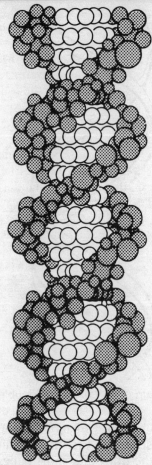

Figure 7.11 A space filling model of DNA.

and Gosling had also come up with something, before learning of the Watson–Crick model, and 'they have it all written . . . so at least 3 short articles in *Nature*'.

The three short papers duly appeared in the issue of *Nature* dated 25 April 1953. The first oddity is that the Watson and Crick paper appeared first, the theory being presented ahead of the experimental data which it explained. But that peculiarity is accounted for by the other oddities in the paper, which presents the model as if inspired by the Chargaff rules, mentions that the published

X-ray data (those of Astbury and of Wilkins) are inadequate to provide a test of the model (thereby implying that the model sprang, as a stroke of inspiration, from Watson's and Crick's basic knowledge of chemistry, not from any detailed X-ray data), and points to the following papers for a more rigorous 'test' of the model – a model actually constructed *using* the King's data. The second of the three contributions came from Wilkins, A. R. Stokes and H. R. Wilson, and presented the general evidence from the X-ray data in support of a helical structure for DNA; the third, signed by Franklin and Gosling, included the vital photograph of the diffraction pattern from type B DNA and confirmed the validity of the Watson–Crick model. Nobody could have guessed from this presentation that the type B photograph had actually been the inspiration for Watson's final attack on the problem.

Franklin never knew just how much Watson and Crick had depended on her data. And until 1969, following the publication of *The Double Helix*, neither Watson nor Crick appreciated just how close Franklin had been to solving the puzzle herself. For the paper which appeared alongside theirs in *Nature*, confirming their model, was only slightly different from the draft which Wilkins had seen *before* receiving the draft of the Watson–Crick paper. The draft, dated 17 March 1953, was discovered among Franklin's papers as a result of the interest roused by publication of Watson's book. It details the structure of DNA as a double helix, although, of course, it does not include the specific idea of base pairing, and had it been published in its original form in the spring of 1953, and Watson and Crick had not by then had the good fortune to have learned about the tautomeric structures of the key bases from Donohue, it would certainly have made a major impact in molecular biology, and the whole story might have emerged in more logical fashion.* It might put Franklin's work in perspective to say that her structure for DNA, established in March 1953, was a lot closer to the truth than Linus Pauling's pro-

* See Sayre, pages 163 and 164.

posed structure, published at about the same time.

But those who take up Franklin's case on her behalf, posthumously, should remember that she never felt the need to take up the case herself. Her paper with Gosling, in draft on 17 March and modified to take account of the Watson–Crick model, was designed as her farewell to both DNA and to King's College. She was happy to be leaving King's for the more congenial atmosphere (as far as she was concerned) at Birkbeck and she was happy to see the back of the DNA work which had been associated with the uncongenial atmosphere at King's. Watson tells how her 'instant acceptance of our model at first amazed me',* not realising that she and Gosling were already halfway to the same model, and not appreciating her happiness at leaving King's, which may in itself have put her in a magnanimous mood. Her earlier 'unconcealed hostility' was replaced by 'conversation between equals', and she discussed her data with Crick with 'obvious pleasure'. 'For the first time he was able to see how foolproof was her assertion that the sugar-phosphate backbone was on the outside of the molecule. Her past uncompromising statements on this matter thus reflected first-rate science.'

Franklin's real tragedy, indeed, is simply that she died young. For there can be no doubt that a Nobel Committee that was sufficiently well-informed and flexible in its interpretation of the rules to give the 1962 prize in Physiology or Medicine jointly to Crick, Watson and Wilkins, while giving the prize in Chemistry in the same year to Kendrew and Perutz for their work on proteins, would also have found room to honour Franklin, even if giving the prize jointly to four people would have created a precedent. One inflexible rule of the awards, however, is that they are only given to the living.

# Ultracentrifugal proof

By the spring of 1953, the evidence in favour of the double

---

* Quotations in this paragraph from *The Double Helix*, page 124.

helix was overwhelming. Few molecular biologists doubted the accuracy of the model, and publication of the Watson–Crick idea, first in *Nature* and then in more detail elsewhere, inspired many researchers to investigate the nature of the genetic code and the puzzle of how the long double helices which make up the genetic material in chromosomes actually manage to unravel and replicate during cell division, as well as how they manage to pass on messages to the cell to enable it to construct proteins. The discovery that the molecule of life is a double helix was a new beginning in molecular biology, not an end in itself. But before we look at the new world of understanding which lies beyond the discovery of the double helix, the story of the double helix itself can be rounded off rather neatly by an account of the experiment which *proved* the nature of its replication. Although the double helix captured the imagination of molecular biologists in 1953, it took another four years to come up with experimental proof of the reality of the nature of this self-replicating molecule.

There are two outstanding features of the Watson–Crick model of DNA as a double helix. First, any sequence of base pairs along the helix can be accommodated by the structure. Looking at just one strand of the helix – either one – there is no restriction at all on the sequence of the four base 'letters' A, C, G and T. The complementary strand mirrors whatever permutation of letters the other strand carries, so that it is the sequence along one strand that matters, and there is complete freedom for any 'message' to be written in the four letter code. This is more restrictive than the 20 letter polypeptide code (but not embarrassingly so, as we have seen) and certainly far less restrictive than the good old Morse code, invoked by Schrödinger as the archetypal simple code in his book, where he first presented the idea of this kind of genetic code to a wide audience. The story of how the code was cracked is told in the next chapter.

The second striking feature of the Watson–Crick model is the way it provides for easy replication. Before this idea

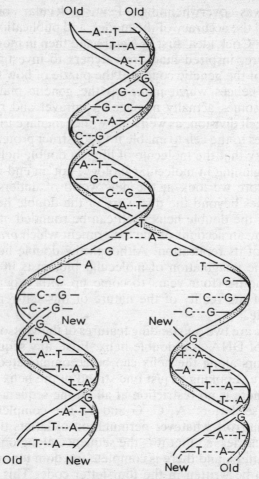

Figure 7.12 Watson and Crick realised that DNA could make perfect copies of itself. As the two strands of the double helix unwind, each pairs up with the appropriate bases and a growing sugar-phosphate backbone to form a new double helix. The two new helices are identical to each other and to the original – each has the same bases strung along its spiral in the same order. So each carries the same message written in the genetic code.

was put forward, it was always imagined that replication must involve a two-stage process. First, the molecule would have to create an intermediate form, a 'negative', comparable to the way in which we might make a mould by pressing a coin into Plasticine. Then the negative, or mould, would have to be used to create a new molecule, equivalent to pouring wax into the mould to create a positive image of the coin. The double-stranded structure of DNA, with complementary bases paired along the length of the molecule, meant that it carried its own 'negative' and cut out one stage in the hypothetical replication cycle. Provided that the chain could unwind with each strand remaining intact, the chemical machinery of the cell would, surely, have no difficulty in assembling a new chain to partner each of the single strands, pairing up the appropriate base at each point along each single chain (A with T, C with G) to create a new complementary strand, producing two double strands of DNA where only one existed before. The difficulty of explaining, in 1953, just how the long molecules could unwind, replicate and coil up again without getting tangled and broken did not detract from the beauty and simplicity of the concept. All of this was spelled out by Crick and Watson in their later papers, a second one in *Nature* and a complete account which appeared in the *Proceedings of the Royal Society*. In the autumn of 1953, Watson left Cambridge for California, and the famous collaboration was over – but there was to be one parting shot from Watson which now reads with the most delightful irony.

The BBC asked Crick to give a talk for the highbrow Third Programme, as it then was, about the discovery of the nature of DNA. Informed of this plan, Watson wrote from Pasadena to say that he thought this would be the height of bad taste. 'There are still those who think we pirated data and I'm of the belief that a few enemies are worse than a few admirers,' he said, but 'you are the one to suffer most from your attempts at self-publicity', and 'if you need the money that bad, go ahead. Needless to say, I shall not think any higher of you and shall have good reason to avoid any further collaboration with you.'* More than a little rich, coming from the man who was later to offend almost everybody involved in the search for the double helix with his own book, generally regarded in the trade as a successful and lucrative attempt at self-publicity.

This mode of replication has a dramatic implication. Although each double strand of DNA unwinds during mitosis and builds itself a new partner (and how well these new ideas explained the dance of the chromosomes observed during cell division), neither of the original strands is destroyed. In all of the cell divisions that have occurred since conception to create the human being that is you out of a single fertilised egg cell, the original strands of DNA forming the chromosomes from each of your parents have never been destroyed, merely unwound and rewound many, many times. Somewhere among the cells of your body you still carry those original strands of DNA inherited from your parents – not just copies, but the self-same atoms forming the self-same molecules. It's an intriguing thought; and it provides the basis for the test which finally proved that DNA does indeed replicate in this way.

This sort of replication is called semiconservative, because one strand of DNA – half the original molecule – is passed on unchanged to the daughter cells. Two alternatives to this kind of replication could, conceivably, exist. In conservative replication, the whole double stranded molecule would be copied, and one daughter cell would get the copy while the other one got the complete original molecule. Alternatively, in dispersive replication the original

---

* Letter dated 9 October, cited by Judson, page 186. Judson also tells how earlier in 1953 Franklin had been officially 'requested' by Randall not to do any further work on DNA, but nevertheless completed writing up her work with Gosling, producing a series of publications which amply confirmed the Watson–Crick model. Her own work at Birkbeck went well, right up until her death; Crick also went on to an eminent career in research, but Watson never did anything else in the same league as his work with Crick. He wrote a great textbook and became a first rate administrator, heading the Cold Spring Harbor laboratory.

DNA might be broken down into smaller units before being copied, with the original pieces getting shared out more or less evenly among the daughter cells and later generations. It was Matthew Meselson* and Franklin Stahl, working at Woods Hole Marine Biological Laboratory, in Massachusetts, who developed a beautiful technique to distinguish between these possibilities.

Like so many of the best experiments it was conceptually simple but very difficult to carry out. The first step was to find a way to distinguish in principle between the original strands of DNA in a colony of growing cells and the new strands made by building up the constituents from the cell medium – the food the cells needed to grow. The obvious approach was to feed the cells a diet rich in a heavy isotope of one of the essential elements, and after trying several possibilities Meselson and Stahl settled on a heavy form of nitrogen, nitrogen-15. The cells they found best to work with were *E. coli*, the familiar gut bacteria. The second part of the experiment had to measure the differences in density between the DNA molecules of different generations of *E. coli*, and that is where the experimenters came up with something really special.

What they needed was a solution in which they could float DNA. Cesium chloride, like common salt but heavier, proved ideal. Then, they needed a way to produce a gradient of density down a tube containing the salt solution. They used our old friend, the centrifuge, but in the latest, souped up, highspeed form, the ultracentrifuge. Whirled around in the ultracentrifuge at 45 000 rpm, a tube of cesium chloride solution gradually established a density gradient, as the heavy salt molecules were forced towards the bottom of the tube, making the solution denser, and pulled away from the top of the tube, leaving slightly

---

* Meselson's name has been in the news again recently, in a very different connection. He is one of the scientists who carried out the chemical analysis which proved that the so-called 'yellow rain' in South East Asia, claimed by the military to indicate communist use of chemical weapons, is actually bee droppings. See, for example, *Nature*, volume 309, page 205, 17 May 1984.

lighter liquid behind. If the solution also contained a little DNA, this would settle down in a very narrow band of the salt solution, in the region where the density of the liquid exactly matched the density of the DNA molecules. Armed with a camera which could take photographs of the spinning tube in the ultracentrifuge (no mean feat in itself), using ultraviolet light to show up the DNA as a dark band across it, Meselson and Stahl could compare the density, and mass, of different lots of DNA molecules.

When everything was ready, Meselson and Stahl grew a colony of *E. coli* in a medium in which all of the nitrogen was nitrogen-15. After many generations, all of the DNA molecules in all of the bacteria alive in the colony were made of, among other things, heavy nitrogen, and when some of the cells were killed and the DNA analysed by the ultracentrifuge technique it showed up as a dark band low down in the tube. Then, the growing colony of bacteria was taken off its heavy nitrogen diet and fed the ordinary isotope, nitrogen-14. After the time required for one new generation of bacteria to grow, the new bacteria were analysed in the same way. The result was again a single dark band of DNA in the ultracentrifuge tube, at a position corresponding to a density exactly halfway between that of the DNA labelled with heavy nitrogen and the density of ordinary DNA. The explanation could only be that in the new generation of *E. coli* each DNA molecule consisted of one strand of heavy DNA inherited directly from its parent, and one strand of light DNA manufactured out of its food. In the second generation, just as the ideas of semiconservative replication predicted, Meselson and Stahl found two stripes in their tube, corresponding to two types of DNA, one hybrid, like the DNA of all the first generation cells, and the other corresponding to ordinary DNA, produced when the ordinary DNA in hybrid cells had unwound and built itself a new partner from a chemical medium containing only normal nitrogen atoms. The experiment, completed in 1957 and published in 1958, showed exactly the same pattern of inheritance as Mendel's experiments with peas – pairs of things being divided and forming new pairings. But where Mendel was

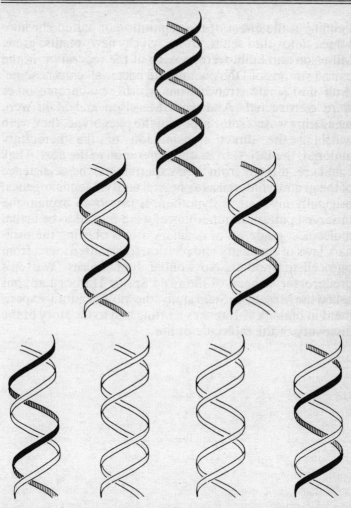

Figure 7.13 Each strand of DNA in the original is conserved during replication. Meselson and Stahl proved this using heavy isotopes to trace the original strands through succeeding generations of DNA. First generation daughters each have one heavy DNA strand; half the second generation daughters have a heavy strand, the other half are all composed of ordinary isotopes.

looking at the effect of the separation of paired chromosomes into the seeds from which new plants grew, Meselson and Stahl were looking at the separation of the paired strands of DNA within the bacterial 'chromosome' itself into single strands from which new chromosomes were constructed. And what Meselson and Stahl were measuring wasn't some aspect of the phenotype, they were watching the direct transmission of the hereditary material, the DNA, from one generation to the next. They could see, directly from their experiments, the persistence of the genetic material as a physical entity. Mendel's genes, originally invoked as hypothetical factors to explain the observed pattern of inheritance, were proved to be actual molecules, made up of ordinary atoms obeying the ordinary laws of chemistry and physics, transmitted intact from one cell to another. No wonder John Cairns, Watson's predecessor as head of the Cold Spring Harbor Lab, has called the Meselson–Stahl study 'the most beautiful experiment in biology'.* It makes a fitting end to the story of the discovery of the molecule of life.

* Judson, page 188.

# PART THREE

# . . . AND BEYOND

'We're going to have a completely new
understanding of the relationship of DNA, the cell,
and the organism as a whole.'

Barbara McClintock
*Science 81*, October

# PART THREE

# AND BEYOND

# CHAPTER EIGHT

# CRACKING THE CODE

It was a physicist, Erwin Schrödinger, who published the first properly thought out idea of a genetic code in the form that it is now understood. His biochemistry was wrong – when he wrote *What is Life?* he thought that protein molecules carried the code of life. But the idea of a code analogous to the Morse code has stuck. And it was another physicist, George Gamow, who brought the idea forcibly to the attention of molecular biologists in the early 1950s, just after Watson and Crick had published their first two papers on the nature of DNA.

Gamow was an ebullient character, born in Russia in 1904 and a graduate of the University of Leningrad, who had worked in Göttingen and at Niels Bohr's institute in Copenhagen with the founding fathers of quantum physics. He moved to the United States in 1933, where his main lines of research concerned nuclear physics and the origin of the Universe. His legendary sense of humour led him into escapades which don't fit in with the conventional image of scientists as respectable, and dull, stuffed shirts, but perhaps give a more accurate insight into how some scientists, at least, work. When, in 1948, his research with a colleague called Ralph Alpher resulted in a major scientific paper about the Big Bang origin of the Universe, Gamow impishly added the name of his friend Hans Bethe to the paper, even though Bethe had made no contribution to the work. Gamow's intention was that the paper would for ever be known as the 'Alpher, Bethe, Gamow' paper, or '$\alpha\beta\gamma$', and his wish was fulfilled.* Later, Gamow became famous outside scientific circles, as the author of a series of books explaining science to the layman; his most memorable creation was 'Mr Tompkins', a character who shrank to the size of an atom and experienced quantum effects first hand. Unfortunately, an attempt by Gamow to get one of his papers published under the names of Gamow and Tompkins was squashed by an editor who had a broad enough knowledge to appreciate that Mr Tompkins was a fictional character, but lacked the appropriate sense of humour to go along with Gamow's little joke.

In 1953, Gamow was visiting the Berkeley campus of the University of California and, as he later recalled,

> I was walking through the corridor in Radiation Lab, and there was Luis Alvarez going with *Nature* in his hand . . . he said 'Look, what a wonderful article Watson and Crick have written.' This was the first time that I saw it. And then I returned to Washington and started thinking about it.'†

Gamow soon wrote to Watson and Crick, introducing himself and presenting the first fruits of his thoughts about the coding problem. These ideas were published in *Nature* in February 1954. He suggested that protein molecules were built up directly on the DNA, which acted as a template which lined up amino acids in the right order. The

---

* The paper is still referred to today, as one of the key works which led to the idea of the cosmic background radiation, the 'echo' of the fireball of creation, the Big Bang itself. And, as often as not, it is indeed simply referred to as the $\alpha\beta\gamma$ paper. To Gamow's delight, purely by coincidence the paper was published on 1 April.

† From an interview in the George Gamow Collection at the Library of Congress, Washington. Quoted by Portugal and Cohen, *A Century of DNA*, page 285. The issue of *Nature* referred to was the one dated 30 May, with the second of the Watson–Crick papers in. Incidentally, this is the same Luis Alvarez who became famous in the 1980s for his espousal of the idea that the death of the dinosaurs, 65 million years ago, was caused by the impact of a giant meteorite with the Earth.

idea depended on Gamow's speculation that the arrangement of the bases along the double helix produced a series of holes with slightly different shapes, the exact structure of each hole depending on which of the four bases made up the sides of the hole. He thought that individual amino acids might plug in to specific holes, and that when all of the constituents of a polypeptide chain had been plugged into a stretch of DNA in this way they would link up with one another, unplug from the holes, and form a complete protein molecule. The idea was full of bad biology. It didn't take a great deal of investigation to show that the gaps between the bases in the double helix are not sufficiently different from one another for amino acids to be selected by plugging them in to specific holes, and Gamow's code soon ran into trouble because of the severe restrictions this kind of model placed on which bases, and hence which amino acids, were 'allowed' to sit next to each other. But Gamow's work was crucially important because, coming hot on the heels of the double helix papers, it forced researchers such as Crick to concentrate immediately on the coding problem, the puzzle of how a string of bases along the double helix became translated into a string of amino acids along a protein.

As Gamow put it in his *Nature* paper, 'the hereditary properties of any given organism could be characterised by a long number written in a four-digital system. On the other hand, the enzymes, the composition of which must be completely determined by the deoxyribonucleic acid molecule, are long peptide chains formed by about twenty different kinds of amino acids, and can be considered as long "words" based on a 20-letter alphabet. The question arises about the way in which four-digital numbers can be translated into such "words".'*

The answer to that question could only be obtained with the aid of a lot more thought about the coding idea in general, plus a far better grasp of the biochemical realities than Gamow had in 1953.

---

* *Nature*, volume 173, page 318, 1954.

# The other nucleic acid

DNA is not the only nucleic acid in the cell. Indeed, just before the publication of the Watson–Crick papers on the double helix, many people thought that the other nucleic acid, RNA, played the more important role in the life of the cell. The early work on RNA is confusing – the researchers didn't really know what they were looking at, or what part it played in the processes of life – and a great deal of the early research was carried out in European laboratories during the Second World War, published (if at all) in different languages and circulated with great difficulty. So the best place to pick up the threads of the story is in Cambridge, in July 1946, when a symposium organised by the Society for Experimental Biology provided the chance to bring everyone up to date before the new wave of research which began in the post-war years.

The important points, in those years, seemed to be a contrast between the stability of DNA in the cell and the variability of RNA. The amount of DNA is constant, and it is the same in all the cells of an organism as well as being the same all the time in each cell. What's more, it stays in the nucleus, out of the way of all the busy activity of the biochemical factories operating in the cytoplasm which makes up the bulk of a cell. RNA, on the other hand, shows a lot of variation. Most significantly, a growing cell contains more RNA than a resting cell, and an active cell, one from an organ such as the liver which is busily manufacturing proteins, is also particularly rich in RNA. And, unlike DNA, RNA could be found in the same place that protein was being manufactured, in the cytoplasm, where it appears in the form of tiny, round structures, now called ribosomes, rich in both RNA and protein. All of the evidence suggested that RNA was directly involved in the manufacture of the enzymes that are vital to the functioning of a cell or a multicellular organism. But how does the RNA 'know' which proteins to make? Equally, even in the 1940s the evidence was accumulating that DNA is the storehouse of genetic information. Although the idea didn't immediately take the world of biology by storm, the key concept appeared in the English summary of a paper by André Boivin and Roger Vendrely, working in Strasbourg, in the journal *Experientia* on 15 January 1947. An unknown editor translated the heart of their paper as indicating that DNA makes RNA, and RNA makes protein. This concept is now at the heart of the understanding of the workings of the cell, and of the genetic code, which developed during the 1950s and 1960s. It pulls the rug from under ideas, like Gamow's first stab at the coding problem, which see DNA manufacturing protein directly, without the aid of an intermediary, and it places RNA very much at the centre of the stage.

At about this time it also became clear that a close relation to RNA plays another important role in the life of the cell. Fritz Lipmann, a biochemist who worked in Copenhagen and then moved to Cornell and finally to the Massachusetts General Hospital, spent much of the late 1930s and early 1940s investigating where cells get the energy they need to operate their 'factories' which stick amino acids together. Lipmann knew, from earlier researchers, that energy is carried around in the body by small molecules rich in phosphorus, adenine triphosphate, or ATP. ATP is a lot like the fundamental unit of RNA. It has a ribose ring, with a base – adenine – on one end, and a phosphate group on the other. But instead of the phosphate linking up to another ribose, it has two more phosphate groups attached to it. Such a molecule is rich in energy – it is made by a process which depends, ultimately, on the input of energy from sunlight during photosynthesis.* The bonds between the three phosphate groups are easily broken, giving up energy which is used to drive the muscles in your body. But it also takes energy to make the components of an RNA chain, or a DNA chain, or a polypeptide chain, join together. Lipmann realised that any base, not just adenine, could be part of a phosphate-

---

* Animals, of course, steal their energy-rich molecules from plants, or other animals, by eating them.

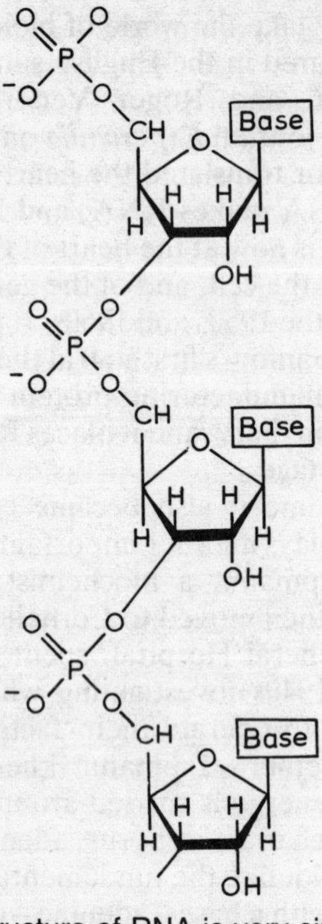

Figure 8.1 The structure of RNA is very similar to that of DNA (see Figure 7.6).

rich molecule like ATP, and that the sugar in such a molecule could just as easily be deoxyribose as ribose. The components from which the cell builds a molecule such as RNA are not the naked bases themselves, but the nucleotides with extra phosphates attached, carrying their own supply of energy which is sufficient to overcome the energy barrier which would otherwise prevent them from sticking on to the growing end of the chain. The phosphates are thrown away and recycled by the cell. A similar process operates to provide the energy needed to join the amino acids together in peptide chains. Lipmann received the Nobel Prize for his work in 1953.

So by 1954 the pieces of the puzzle were falling into place. The evidence that DNA carries the genetic information had become compelling; the structure of DNA was known to be a double helix, a structure which obviously lent itself to copying; the source of the energy used to join long chains of biomolecules together had been identified; and even the idea that DNA makes RNA which makes protein was in the air. The problem was how the DNA made RNA, and how the RNA made protein. And the central figure in the attack on that problem was Francis Crick, stimulated in the early stages, at least, by Gamow's prompting. In his characteristic enthusiasm, Gamow invented an 'RNA Tie Club' with twenty members (one for each amino acid) and Fritz Lipmann as an honorary member; the idea behind this was to generate discussion and an interplay of ideas through correspondence, leading to a more or less concerted attack on the coding problem. It never really worked like that, although Gamow's guiding spirit did keep the enthusiasm going long enough to generate some publications. Crick kept in touch with Gamow, and made some 'Tie Club' contributions, and he also kept in touch with Watson, now back in the US. But there was also now another member of the inner circle, a South African, Sydney Brenner, who obtained a PhD from Oxford University in 1954 and then worked for a time in South Africa before returning to England to join the MRC's Molecular Biology Laboratory in Cambridge in 1957. Crick's letters to Brenner in South Africa, discussed in detail by Horace Judson in *The Eighth Day of Creation*, provide the best insight into how the attack on the genetic code developed in the mid-1950s.

# Triplets to the fore

Whether you think in terms of RNA or DNA the principle of the coding problem is the same. Each nucleic acid is made up of four bases, and the fact that the RNA bases are UCG and A while the DNA bases are TCG and A doesn't affect the theoretical understanding of how a four-letter code provides the information needed to write words in a 20-letter alphabet. From now on, I shall generally use the RNA bases when describing the code at work, and how it was cracked, since it did indeed turn out that it is RNA which directly controls the manufacture of protein, having itself been made on a DNA template. Very early on in the game the experts concentrated on a triplet code, in which three bases together (such as UCG) correspond to one type of amino acid, to one letter in the polypeptide alphabet. The reason is simple. If you take just one base at a time, then you can only 'code' for four different things. But if you take two bases at a time, in a doublet code, you can make 16 different arrangements of paired bases (4 × 4), such as UC, because whichever one of the four you take as the first member of the pair, you have a choice of four different bases for the second member of the pair. Sixteen combinations, however, still isn't enough to code for the 20 different amino acids that are fundamental constituents of all living things. If you take the bases three at a time, in triplets, you get 64 different combinations (4 × 4 × 4), and that is more than enough to code for all 20 amino acids, with some combinations left over to use as 'punctuation', codes which indicate the beginning and end of an amino acid chain – 'start' and 'stop' codes, as far as the enzymes which build the chains are concerned.

The ideas building up to an understanding of the triplet code developed slowly in the 1950s. As early as 1952, Alexander Dounce, a biochemist working at the University of Rochester, published a paper which discussed the possibility of RNA acting as the template which controlled the manufacture of proteins, and which included the idea of a code which depended on the arrangement of RNA bases in threes to identify the specific amino acids in the protein chain. Even then, to a researcher such as Dounce, it seemed obvious that the RNA itself was copied, somehow, from the DNA in the nucleus. The paper was ahead of its time – published a year *before* Watson and Crick described the double helix – and made no great impact, although it is looked back on as a landmark, the first clear presentation in print of the idea that the exact order of the amino acids in the protein depends on the exact order of the bases along the nucleic acid chains. This was the idea that Crick championed throughout the decade following the discovery of the double helix.

Different variations on the theme were tried, and found wanting, such as the idea that the code might be an overlapping one in which you get a different, but meaningful, message if you start reading from the second 'letter' of a triplet instead of the first. But ultimately the theorists were forced back on the simplest possibility, that the string of bases along a nucleic acid have to be taken three at a time, starting in a definite place and stopping in a definite place. They can spell out a message such as UCG TCG TCU GGC CCT in which each triplet corresponds to one amino acid. Such a code has some striking features which immediately give insight into the whole business of mutation, the mechanism which provides the variation that is the key to evolution by natural selection. If you take out one of the bases, then the whole of the message from that point onward is hopelessly scrambled, assuming that the cell's reading mechanisms carry on translating the code in triplets. A message such as JIM GOT THE HOT PIE would become, simply by taking out one letter, JIM GTT HEH OTP IE. Similarly, just by inserting one extra letter – one extra base – we would get the equally nonsensical JJI MGO TTH EHO TPI E. But if you substitute one letter, or one base, for another then you change the meaning of only one word in the message – one amino acid in the polypeptide being manufactured in accordance with the coded instructions. That might make a nonsense word in the middle of the message, such as JIM GOT QHE HOT PIE. Or it might introduce a new, but plausible, word into the message: JIM GOT THE HOT PIG. In that case, if the theory was correct, the cell's factories would take the information and use it to construct a polypeptide chain differing in one amino acid from the composition that it ought to have, perhaps putting a *val* where there ought to be an *ala*. Such a modified protein might work perfectly well at its job. Conceivably, it might even do the job better than the original, giving the organism which carried the mutated form of the genetic code for manufacture of that particular protein a slight edge in the struggle for survival. Or it might prove fatal, if the result was a protein that could not perform the function for which it was required. The discovery of just such a mutation at work, producing the disease known as sickle cell anemia, persuaded the biological world that the code breakers really might be on the right track.

## The sickle cell clue

In the 1940s Linus Pauling learned about sickle cell anemia through his work on a committee set up to decide which areas of medical research the US Government should sponsor after the war. Sickle cell is a disease in which the red blood cells, which contain hemoglobin and carry oxygen around in the blood, become distorted, crumpling into a sickle shape which gives the disease its name and makes them ineffective. In extreme cases, the disease causes death. But it is especially interesting to geneticists because of the way the gene for sickling survives among the gene pool of West African negroes. The same mutation which causes sickling can provide protection against malaria, so in regions where malaria is endemic there is a trade off, sufficient to ensure that the recessive gene stays in the gene pool.* The sickle cell disease only strikes if an individual inherits the gene from both parents, and in evolutionary terms this risk has proved worth the benefits of the protec-

---

* It happens like this. People who carry the sickle cell gene and the normal gene as a heterozygous pair have hemoglobin which has a slight, but not severe, tendency to sickling under normal conditions. If they are attacked by the malaria parasite, the parasite gets inside the red blood cells, just as it does with people who don't carry the sickling gene. But the red blood cells from the person with sickling trait crumple up when they are invaded in this way, and the invaders are destroyed along with them. So, if you are in a region where malaria is common, it pays to carry the sickling trait in the genetic code of one gene in a heterozygous pair, even though those of your children that have the sickling trait on a homozygous pair may die.

tion against malaria – but this isn't so, of course, among the descendants of those West Africans, now living in the United States, where malaria is less of a problem.

Pauling's interest in sickle cell concerned the chemistry of the sickling phenomenon. He surmised that there must be a chemical difference in the hemoglobin of sickle cell patients compared with normal hemoglobin, and that this difference caused the change in shape of the cells. He got one of his students, Harvey Itano, to test the idea by trying to find a difference between hemoglobin from normal individuals and from sickle cell sufferers, and after some painstaking experiments Itano discovered that hemoglobin from normal people carries a slight negative electric charge, while hemoglobin from sickle cell patients has a slight positive charge. The proof that sickle cell anemia is what Pauling called 'a molecular disease' was published in 1949, the same year that James Neel, of the University of Michigan, published a paper which established once and for all that the disease is caused by a mutant, recessive gene, that is transmitted from one generation to the next exactly in line with the Mendelian laws of inheritance. Together, the two pieces of work showed that there is a specific chemical change in the hemoglobin of sickle cell patients, a chemical change produced by a change in a single gene carried in the human chromosome set. That discovery, linking Mendelian genetics, Darwinian evolution and biochemistry, was striking enough. But more was to come.

Crick and the rest of the inner circle of code breakers knew about the sickle cell experiments, and, deeply immersed in their thinking about DNA, RNA and the genetic code, they took it for granted that the biochemical change in the protein molecule produced by the mutant gene was a change in the amino acid composition. But it *could* have been that the change in the visible charge was simply due to a change in the way the hemoglobin chain was folded, exposing a different portion of the chain to chemical view. The proof that a change in the amino acid composition of the chain was involved came from a study carried out at the Cavendish by Vernon Ingram in the mid-1950s.

Fred Sanger, who also worked in Cambridge, had completed his analysis of the structure of insulin by 1955. Both the principle and the experimental techniques for sequencing a protein into its constituent amino acids had been established. Ingram, trying to find out just what it was that made the hemoglobin in sickle cell patients different from normal hemoglobin, was faced with a slightly different problem from the one tackled by Sanger, but a closely related one. He wasn't trying to determine the complete sequence of amino acids in either version of hemoglobin; what he had to do was find the *difference* between two polypeptides. So his task was first to identify a stretch of this particular polypeptide that was different in the two kinds of hemoglobin, and then to analyse in detail, using much the same techniques as Sanger, just this short stretch of material of special interest.

In the first stage of the experiment Ingram used an enzyme called trypsin to chop the hemoglobin chains into chunks that were more manageable, chemically speaking. Trypsin breaks the peptide bonds selectively, always next to either lysine or arginine, and only on one side of each of those two molecules. Because trypsin always cuts the chain in the same places, Ingram could be confident that every time he prepared a sample in this way he would obtain exactly the same fragments – about thirty stretches of polypeptide each containing only about ten amino acids. What was a very small chemical difference between two chains, roughly 300 amino acids long, ought to show up as a much bigger proportional difference between two of these short, ten-acid chains. And that is just what he found. Ingram used a mixture of chromatography and electrophoresis to separate out the fragments produced by breaking up first normal and then sickle cell hemoglobin. The electric potential was applied at right angles to the chromatographic separation, so that as the different polypeptide chunks migrated up through the paper those with a negative charge were pulled to one side, those with a positive charge to the other, while the neutral fragments went straight up. When the papers were dried and stained in the usual way to show up each fragment of chain as a coloured blob on the filter paper, Ingram found what he was looking for. On the chromatogram from sickle cell hemoglobin, the identifying 'fingerprint' of the molecule, there was one extra blob on the side from the neutral line corresponding to positive charge. Comparing the chromatograms carefully, Ingram found that this corresponded to a neutral blob in the normal hemoglobin chart. He had pinned down the chemical difference between the two hemoglobins, and confirmed that it involved exactly the charge difference that Pauling's team had found.

The next step was to dissect the interesting blob chemically and determine its complete amino acid sequence. Using the same techniques developed by Sanger, it was relatively simple to sequence a peptide chain containing, as it turned out, just eight amino acids, and to compare this with the sequence of the corresponding stretch of chain from normal hemoglobin – the neutral blob on the other chromatogram. The tests established that the difference between normal and sickle cell hemoglobin consists of a change in just one amino acid. A glutamic acid present in normal hemoglobin is replaced by a valine at the same site in the polypeptide chain of sickle-cell hemoglobin. Glutamic acid has an extra acidic group in its structure, compared with valine, and this group carries the extra negative charge which makes the fragment as a whole electrically neutral under the conditions prevailing in the cell. The absence of that negative charge leaves a positive balance, the difference that had shown up during the fingerprinting of the chains, a difference that can prove fatal in individuals that carry the sickling gene on both chromosomes. The discovery was announced to a scientific meeting in London in September 1956, and published in *Nature* in October. It made an enormous impact, because it proved what the inner circle of code breakers already knew, that a genetic mutation inherited in Mendelian fashion corresponded to a change – in this case, the simplest possible change – in the amino acid sequence of a protein. And the developing idea of a triplet code explains just how such a mutation can occur. Suppose that, in the long sequence of nucleic acid bases – the gene – coding for hemoglobin, one triplet, GGA, codes for glutamic acid, while another, GUA, codes for valine. A copying error which resulted in a change of just one base out of a whole DNA molecule would lead to the production of RNA with a U where there ought to be a G, and that single change in the code, a point mutation, would result in the production of sickle-cell hemoglobin instead of normal hemoglobin. In an interview with Judson, Crick later stressed the importance of this breakthrough:

> It was a perfect case . . . Ingram's discovery made an *enormous* difference. Because people suddenly realised that there was this connection. It now became obvious; and what was clear to us before it was done then became clear to everybody.*

But puzzles still remained. How did the genetic message get out of the nuclear DNA and into the cytoplasm? How did the cell's factories piece together amino acids to make proteins? And exactly which triplets of bases coded for which amino acids?

## Adaptors and messengers

Francis Crick had been instrumental in persuading Ingram to take on the task of identifying the difference between sickle-cell and normal hemoglobin. The stimulus came from Paris, where Boris Ephrussi had pointed out to Crick, in the spring of 1955, that there was no actual *proof*, from any studies so far carried out, that a change in a gene caused a specific change in a protein. Plenty of circumstantial evidence, to be sure; but no definite proof. So it was thanks to the Paris group, and to Crick, that Ingram established once and for all that a single gene really does code for a single protein, the 'one gene, one protein' concept that is central to biology today. And the next

* Eighth Day, page 308.

steps forward in the understanding of how the message in the gene is translated into the protein that it is coded for were also made by Francis Crick and researchers in Paris, at the Pasteur Institute.

Crick's next contribution was contained in a paper which he presented to the Society for Experimental Biology in September 1957. It has been described as 'one of the most stimulating and liberating conceptual papers in genetics', and it is well worth reading today.* In spite of its title, 'On Protein Synthesis', this is not a paper for biochemists, but a general view of the problem, including the coding problem, presented in clear language intelligible to anyone interested in the code of life. Crick both summed up the state of knowledge, as of 1957, for a wide audience, and also pointed the way ahead, stimulating others to follow up his leads. He spelled out that 'the main function of proteins is to act as enzymes', and that 'the main function of the genetic material is to control (not necessarily directly) the synthesis of proteins', and he explained the nature of proteins as long polypeptide chains. He described the 'magic twenty' amino acids that go into those protein chains, pointed out the similarities between human and horse hemoglobin, and emphasised the crucial significance of Ingram's work in establishing that 'the gene does in fact alter the amino acid sequence'. Summing up, he said 'the unique feature of protein synthesis is that only a single standard set of twenty amino acids can be incorporated, and that for any particular protein *the amino acids must be joined up in the right order*. It is this problem, the problem of "sequentialization", which is the crux of the matter.' And then he went on to offer his own ideas on how the amino acids are put together in the right sequence.

At that time, RNA was known to be associated with protein synthesis, because cells actively producing protein

---

* The assessment of the importance of Crick's paper is from Elof Carlson, *The Gene*, page 236. Crick's paper was published in the *Symposium of the Society for Experimental Biology*, volume 12, page 138, 1958. Other quotations in this paragraph are from that paper.

were known to be rich in RNA. The only known site of RNA in the cell was in the round particles, the ribosomes, in the cytoplasm, so it was natural for Crick to assume that the RNA templates on which the proteins were constructed were carried by the ribosomes. But how did the arrangement of bases along the template – the RNA – enable the cell to align amino acids in the correct sequence before joining them together to make a protein chain? Crick had hit on the idea, some time before, that there must be small molecules within the cell that carry each amino acid to the template. Such a molecule, which he called an 'adaptor', could have a triplet of RNA bases on one end, which it could match up with the corresponding triplet of bases on the template, and it would have a specific amino acid, corresponding to that particular word in the genetic code, attached to its other end. The adaptors, in other words, were likely to be modified forms of RNA 're-membering' one triplet code word and seeking to deposit their amino acid burden – a single amino acid residue – alongside that code word in an RNA template. As we shall see, just such a family of molecules was indeed soon identified, although Crick's suggested name hasn't stuck, and these molecules are generally referred to as 'transfer RNA', or tRNA.

So Crick suggested that there must be at least two types of RNA in the cytoplasm. He stressed the idea that 'the specificity of a piece of nucleic acid is expressed solely by the sequence of its bases', and that 'once "information" has passed into protein *it cannot get out again*'. This was an idea that had been floating around for a while, but had never been explicitly stated before. 'Information', in this sense, is the precise determination of a sequence, either of bases in the nucleic acid or amino acids in the protein, and what Crick was saying was that although DNA could control the manufacture of RNA, and RNA could control the manufacture of protein, the process could not flow in reverse, using the information in protein to create copies of the RNA and DNA molecules. This fundamental concept he called the 'Central Dogma' of molecular biology.*

---

* Quotations from Crick, op. cit.

Meanwhile, over in Paris, another group of researchers had been studying the mystery of life from a different direction. The *patron* of the group at the Pasteur Institute was André Lwoff, who had been born in 1902, of Russian–Polish parentage; the other key figures in the development of the work there were Jacques Monod, who was born in 1910 and gained his doctorate in 1941, and François Jacob, whose education was interrupted by military service during World War Two and who completed his medical degree in 1947, only joining the Pasteur Institute team in 1950, as a research assistant. Lwoff and Monod were also active during the war, high up in the French resistance movement; all three were making a fresh start in research. The main line of the research was phage, and the way it could hijack the genetic machinery of a bacterium to make the cell produce copies of the invading virus, instead of going about its normal business. But this was just one part of the story. Another concerned sex amongst the bacteria.

In the early 1950s, several groups identified and isolated strains of *E. coli* that often went in for a form of sexual reproduction, instead of the usual asexual cell division that is characteristic of single-celled organisms. This meant that it would be possible to do experiments involving genetic recombination on these organisms, which have a very rapid life cycle, far quicker than even the *Drosophila* which have for so long been the favourites of geneticists. In 1955 Jacob and a colleague, Élie Wollman, began to investigate the way genetic information was transferred from one bacterium to another. In particular, they wanted to identify the location on the bacterium's single chromosome of the genetic information that had been placed there by an invading phage – for Lwoff had been investigating the surprising discovery that a bacterium invaded in this way doesn't always immediately produce a flood of new phages and burst open, or lyse. Sometimes, the genetic material of the phage lies dormant, unless it is activated in some way, quietly reproducing itself along with the host's genetic material every time the DNA is copied and the cell divides. This type of viral infection is called lysogeny, and the virus is called a provirus, or prophage. The puzzle of where the hidden genetic message was located could, in principle, be tackled in the same way as genes for red or white eyes were identified in *Drosophila*, by repeated matings and analysis of the inherited characteristics of the offspring with recombined genetic material. This was particularly straightforward in the case of the hidden phage gene in the bacterium, since this could be activated by ultraviolet light, causing the cell to release its flood of new phage and identifying it as a carrier of the sleeping gene.

Wollmann and Jacob found a simple way to measure off the genes on the bacterial chromosome. They took strains of 'female' *E. coli* (the ones that could receive DNA from a partner) that had certain characteristic mutations, and mixed them with 'male' *E. coli* (the variety that gives up its DNA to a partner, given the opportunity), so that the transfer of genetic material could begin. Then, at regular intervals they took some of the sample of mixed bacteria and examined them to see how effectively the normal genetic information had been transferred into the mutant strain. Their first great discovery was that it could take two hours for all the genetic information to be transferred from one bacterium to another, even though this was a species that normally divided every twenty minutes. The chromosome from the male partner was being slowly squeezed out, like toothpaste from a tube, into the female partner. And the way this showed up provided the second great discovery.

Suppose that in one such experiment the mutant female lacked effective genes for four different properties, A, B, C, and D. By stopping the bacterial matings after certain well-defined intervals (for obvious reasons, this became known as the *coitus interruptus* experiment), the team found that after ten minutes, say, the gene for property A had been transferred to the females, but none of the others had. After fifteen minutes, both gene A and gene B had been transferred; after 20 minutes the first three genes had been transferred; and only after half an hour had all four genes been transferred. The experiment provided a way to count off the genes along the chromosome, a measure of how much of the chromosome had been transferred before

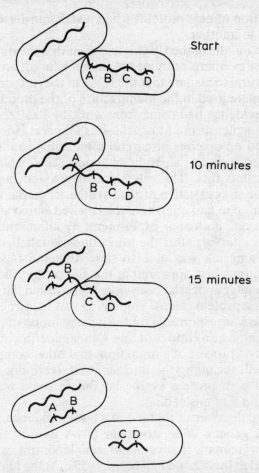

Figure 8.2 The *coitus interruptus* experiment. It takes several minutes for a bacterium to inject its chromosome into another bacterium. By interrupting the process it is possible to count off genes along the chromosome. In this hypothetical example, only two out of four genes have been transferred in the time allowed.

each mating was interrupted. That helped in Jacob and Monod's investigation of how genes are switched on and off, which we shall hear more of in the next chapter. But it also helped in another series of experiments which demonstrated how quickly the new DNA could begin to control the creation of new proteins when it was transferred from one cell to another.

By the middle of the 1950s it was clear that the ribosomes are the sites where proteins are manufactured, and it had been natural to assume that the RNA in the ribosomes was the template used in the manufacture of the proteins. The crucial evidence had come from work by Paul Zamecnik and colleagues at the Massachusetts General Hospital. It depended on a series of careful tests using a radioactive isotope of carbon, $^{14}C$, to label amino acids, which were injected into rats. The rats were then killed, and their livers, the sites of active protein synthesis, mashed up and investigated to find out where the labelled amino acids had gone. In one particular experiment, by killing the rats at different intervals after the injection with labelled amino acids, Zamecnik was able to show that the radioactive carbon atoms first appeared at the site of the ribosomes, and only later appeared elsewhere in the cell in the form of completed protein chains. Something that happened at the ribosome was responsible for the incorporation of the individual amino acids into proteins. Other experiments established, by a process of elimination, that other components of the cell, including the nucleus itself, were not directly involved with protein synthesis. But in some ways that conclusion was misleading.

Crucial discoveries that led to a better understanding of how the genetic code stored in DNA is used to make protein were made in several different laboratories around the world in the second half of the 1950s. At the Oak Ridge National Laboratory in Tennessee, Elliot Volkin and Lazarus Astrachan showed, using a similar kind of radioactive labelling technique, that shortly after a phage invades a bacterium, a small quantity of new RNA is made. Painstaking analysis of the cells established by 1956 that this RNA has a structure very much like a single strand of DNA, a series of bases which mimicked the ratio of the bases in the DNA of the invading phage, not the DNA of the invaded bacterium. In 1959 Sam Weiss, at the Argonne National Laboratories in Chicago, took things a step further still. Starting out with DNA and an extract of the cells of bacteria, Weiss added the four nucleotides which form the bases of RNA to the brew in his test tubes and showed that RNA was indeed manufactured, presumably using the information – the code – in the DNA, and the machinery – the enzymes – from the disrupted cells. And back in Paris Jacob and Monod, working in 1957 and 1958 with Arthur Pardee, a visitor from the University of California, had been puzzling over the speed with which genes that invade a foreign cell can take over the machinery and begin producing the material coded for in their own DNA. This implied that no complex substances with large molecular weight (such as ribosomes) were being made before the synthesis of new proteins could begin, but rather that some very simple messenger was being produced very quickly by the DNA. Since the messenger couldn't be found floating around in large quantities in the cell, it must be being broken up after use. This line of thought led to experiments which showed that, indeed, protein synthesis in the cell stops if the DNA is removed, even though the ribosomes are still there. The factories manufacturing protein could not operate on their own, but only under the guidance of the, as yet, invisible messengers from the DNA.

With hindsight, the implication seems obvious. When a phage invades a bacterium, its DNA immediately and quickly manufactures a form of RNA which is an exact counterpart – apart from the substitution of U for T – of one of the strands of DNA in the helix. This messenger RNA then moves to the ribosomes which belong to the bacterium, and the bacterium's own cells blindly follow the instructions on the RNA, manufacturing the proteins that the phage wants, not the ones that are right for the bacterium. And that must be how ribosomes operate normally, except that under normal conditions they follow instructions contained in the genetic code carried by RNA which has been made by copying the cell's own DNA. Ribosomes are simply idiot machines, blindly manufacturing proteins in accordance with whatever instructions are fed to them, like an automated factory blindly following the instructions on a computer tape. To change the manufacturing process, you don't have to build a new factory, simply change the instruction tape. But in 1960 it still required an act of inspiration to put all of the pieces together and come up with an overall picture of how the cell worked. The inspiration emerged from an informal meeting of several members of the inner circle of code breakers in Sydney Brenner's rooms in Cambridge in April 1960. Crick and Jacob were both there, together with Brenner and two or three other people who had wandered up to Cambridge following a meeting in London of the Microbiological Society. The Cambridge gathering was the catalyst which sparked the full understanding of how the code is used by the cell to make protein, and it led to a definitive paper by Jacob and Monod which pulled all the different bits of information into a coherent whole and, almost as importantly, came up with neat phrases and labels for the mechanisms which made it crystal clear what was going on.

This was the point where Crick and the others, comparing notes and tossing ideas about in uninhibited fashion, at last appreciated that the ribosome was no more than a reading head, and that the information it used came from DNA via a messenger in the form of RNA (the name 'messenger RNA', or mRNA, was one of Jacob and Monod's inventions). Crick later described the failure to appreciate the need for the messenger earlier as being 'the one great howler in molecular biology';* when the group in Cambridge dispersed, they not only spread the news of their new insights to colleagues, but also determined to find the messenger and identify it in the test tube. Jacob and Brenner went to CalTech for the summer, where they talked with Delbrück (who didn't believe them) and Meselson, whose density gradient experiments gave them a means of locating the messenger. Their new experiments involved labelling the components of the cell with heavy isotopes and radioactive atoms, tracking the path of the RNA produced by phage when they infect bacteria. They found what they were looking for.

---

* Judson, page 433.

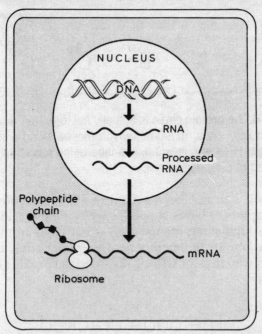

Figure 8.3 Cells like those of our own bodies, eukaryotic cells, carry out several stages of processing in translating DNA in the nucleus into (eventually) protein chains that keep the body working.

Using a different approach, at more or less the same time, a group at Watson's lab in Harvard also found the elusive messenger RNA. Back in Paris in the Autumn, Jacob worked with Monod on the great paper which provided the answers to almost all of the questions Crick had raised in his classic 'On Protein Synthesis'. Theirs was called 'Genetic Regulatory Mechanisms in the Synthesis of Proteins'; it reached the *Journal of Molecular Biology* on 28 December 1960, and was published in 1961. At last, the *mechanisms* of the cell were understood, at least in outline. In 1965 Jacob, Monod and Lwoff received the Nobel Prize in Physiology or Medicine for their contributions. It was the first time a Nobel award had gone to a French scientist for three decades, and it was richly deserved. But in spite of the efforts of Monod, Jacob, Crick and the others still the

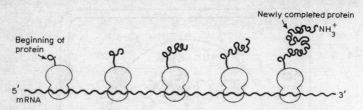

Figure 8.4 The protein chain is actually put together, amino acid by amino acid, by a ribosome, which moves along a strand of messenger RNA and reads it, like tape being read by the magnetic head of a tape recorder.

code itself had not been solved in 1960 – nobody knew which triplets of bases coded for which amino acids. That breakthrough came immediately – but from an outsider, not from one of the members of the inner circle of code breakers who had for so long regarded the puzzle as 'their' problem.

# The code cracked

A book, like the genetic code on the double helix itself, is linear. But it is impossible to tell the history of the double helix in a linear fashion. Like most of science, the story involves different people working in different places but at overlapping times, and the whole cloth can only be seen by weaving the separate threads of the story individually and then stepping back to look at the finished tapestry. Sometimes, it is necessary to take a step or two backwards in time, to pick up a different thread and discover how it relates to the threads that have already been woven into place. So it is with the tale of how the genetic code itself was finally cracked. We have to go back from the feverish excitement of the inner circle in Brenner's rooms in Cambridge in 1960, to work being carried out in a different, but related, context at the New York University School of Medicine in the mid-1950s.

The work was carried out by Marianne Grunberg-Manago, a French biochemist working in the US, and Severo Ochoa, born in Spain but a United States citizen. It involved ATP, the molecule which is so important as an energy carrier for living organisms, and which can be described as a lone nucleotide with two extra phosphate groups stuck on the end. It was the role of ATP as an energy carrier that was being investigated by the group in New York, and as part of this work Grunberg-Manago was 'feeding' enzymes with a supply of ATP to see what use they made of it. One of her experiments didn't turn out at all as expected. Instead of using the energy from the phosphate groups in any of the ways anticipated, one of the enzymes took the ATP and manufactured some peculiar, initially unidentifiable substance. Baffled by this, but appreciating that the unexpected might be at least as interesting as the work she was supposed to be doing, Grunberg-Manago spent months trying to find out what was going on. At last, she identified the mysterious product of the reactions in her test tubes – the enzyme had taken the ATP, discarded the extra phosphate groups from the end, and joined the nucleotides together to make a long chain nucleic acid. It was a very strange nucleic acid, compared with the RNA usually found in the cells – a string of sugars and phosphates joined together in a backbone, as usual, but with the same kind of base attached to each sugar, an endless repetition of the 'message' AAAAAA . . . The enzyme was named polynucleotide phosphorylase, which more or less describes its function, and the simple nucleic acid it manufactured was promptly nicknamed poly-A. And, it was soon discovered, the enzyme was equally happy to take the equivalent molecules containing guanine, uracil or cytidine (GTP, UTP or CTP) and manufacture strings of poly-G, poly-U or poly-C. It would even take a mixture of phosphates which included both adenine and uracil, for example, and string them into a chain of poly-AU, with the different bases scrambled at random along the chain, in no particular order. Nobody knew what the enzyme did in the cell (nobody really knows what it does even today, except that it is *not* the enzyme that joins nucleotides together in the cell). But that didn't stop the biochemists from making use of its remarkable properties.

One such biochemist was Marshall Nirenberg. Born in

New York in 1927, Nirenberg graduated in zoology from the University of Florida in 1948, but moved relatively slowly towards a PhD, which he obtained from the University of Michigan in 1957. Indeed, he was completing his PhD work, at the age of thirty, just at the time when Grunberg-Manago and Ochoa were serendipitously discovering a way to make artificial (but stupid) RNA. With his doctorate freshly under his belt, Nirenberg moved to the National Institutes of Health, in Washington DC, where he soon met up with Johann Matthaei, a German biochemist who came to the United States in the late 1950s on a NATO postdoctoral research fellowship.

Nirenberg and Matthaei became interested in the problem of creating the right conditions for protein synthesis in the test tube, under controlled conditions in which the activity of different cell components could be tested and their roles in the process identified. Such a system, called a 'cell-free' system because it does not contain whole cells (although, of course, it does contain components of cells) derived from the work of people like Zamecnik, whom we have already met, and Alfred Tissieres, at Harvard, who developed a similar system using mashed up *E. coli* instead of rat liver. The NIH team came at the problem from a different angle to the attack then being mounted by Crick, Brenner, Jacob, Monod and their colleagues. As biochemists, they had a different perspective from the start; as slight outsiders, beginners in the art of postdoctoral research, they had the advantage of freshness, not being bound by the 'established wisdom', such as the conviction that ribosomes contained the RNA template, a conviction which held back the thinking of the inner circle at this time. But they knew the published work, and they knew that RNA was an essential component of the mechanism which produced protein. They set out to create a cell-free system in which a message carried by RNA molecules would be translated into protein, in the test tube. And they were even using the term 'messenger RNA', independently of its invention by Jacob and Monod, early in 1961. Nirenberg and Matthaei knew exactly what they were looking for, and they set out on a long, careful series of experiments to find it – there was no serendipity about their work.

First, they set up their cell-free system, based on the design of Tissieres, who passed on useful tips, and got it working. Ribosomes, plus a mixture of nucleic acids, the right conditions of temperature, acidity and so on could be set up in the test tube in such a way that when amino acids were added to the system proteins were manufactured. So far so good. When RNA extracted from ribosomes was added to the system, it made only a marginal difference to its efficiency at making proteins, a slight improvement which only indicated how little information was being carried by the ribosomes. When they added pure RNA from the tobacco mosaic virus (a virus which, logically enough, causes a disease of tobacco plants) the results were much more dramatic, with a rapid boost in protein production. All this was happening early in 1961, and in itself would have been valuable confirmation of the reality and role of messenger RNA. Now, Nirenberg and Matthaei wanted to try the system with artificial RNA, the simplest possible form of the genetic code. By this time, poly-U, poly-A and poly-AU were easily available to anyone involved in this kind of research, and on 22 May 1961 it was the turn of poly-U to be tried as part of the carefully planned-out series of experiments. The results were dramatic. The cell-free system, carefully set up to produce proteins in accordance with whatever RNA instructions it received, promptly started turning out large quantities of phenylalinine in long chains. The idiot nucleic acid carrying the message UUUUU . . . caused the ribosomes in the test tube to manufacture an equally monotonous protein, containing one amino acid, *phe.phe.phe.phe*. . . .The first part of the genetic code had been broken – poly-U translated as poly-*phe*.

The news broke at a meeting in Moscow that summer – the Fifth International Congress of Biochemistry. Nirenberg, a junior researcher with no great reputation, was allotted time on the itinerary to present a ten-minute paper in one of the smaller rooms. Only a very few people had got wind of the importance of the work, which hadn't

yet been published, but among the handful of people in the
audience was Meselson, who was, in his own words
'bowled over by it'.* Meselson went immediately to Crick,
who was the chairman of one of the closing sessions of the
meeting, and told him the news; Crick, in turn, re-arranged
the itinerary of his own session so that Nirenberg could
present his paper again, in the main hall and to the full
audience of conference participants. 'After the presentation', recalls Meselson, 'I ran up to Nirenberg and I
embraced him. And congratulated him . . . it was all very
dramatic.'†

Everyone in the trade rushed away from the Moscow
meeting determined to try the experiment, and the results
were quickly confirmed and extended. In 1968, Nirenberg
received the Nobel Prize, and by then the entire genetic
code had been determined, starting out from his work
which led to the identification of poly-U with poly-*phe*. In
the early summer of 1961, of course, it still wasn't certain
how many Us it took to code for one *phe*, although the best
guess was a triplet, UUU. Fired by the work of Nirenberg
and Matthaei, Crick and Brenner now resolved that puzzle
in double quick time, in a series of experiments brilliant for
the simplicity of their approach. Remember that adding or
removing one letter in the coded message scrambles the
whole of the message from there on, and in terms of the
genetic code such a mutation would make a gene useless,
causing the manufacture of a piece of nonsense protein (or
nothing at all) instead of a useful enzyme. Adding or
deleting *two* letters is just as bad. But now suppose that
*three* letters are either added or subtracted all within a
short stretch of the message. A message such as THE CAT
AND THE RAT SAT PAT might become THE SCA
TAA DND THE RAT SAT PAT, or it might become TEC
AAN THE RAT SAT PAT. In either case, the point is that
although a short stretch of message is garbled, because the
overall effect of the 'mutations' is to remove or add exactly

* Interview with Judson, *Eighth Day*, page 481.
† Ibid.

one three-letter word, the rest of the message still makes some sort of sense. A gene altered in this way might still function, at least partially, because it would lead to the manufacture of a protein in which only a few of the amino acids were altered. That is just what Crick and Brenner, and their colleagues, proved, using mutant strains of phage. The paper announcing their results appeared in the last issue of *Nature* of 1961, under the title 'General Nature of the Genetic Code for Proteins'. It established once and for all that a group of three bases codes for one amino acid, that the code is not overlapping, that the sequence of bases is read from a fixed starting point, and, almost as an aside, that the code must be 'degenerate', since with 64 triplets and only 20 amino acids to code for, some amino acids must be coded for by more than one triplet, or 'codon', as Brenner dubbed the basic unit for the genetic message.

Although it took five years to solve all of the rest of the puzzle and identify exactly which codons code for which amino acids – much of the work being done at Ochoa's lab, synthesising many different kinds of RNA and testing to see which proteins they produce in the cell-free system – the appearance of this paper neatly in the last *Nature* of the year effectively marked the end of the search for the double helix. It is hardly a coincidence that the following year, 1962, saw the award of the Nobel Prize to Crick, Watson and Wilkins (for physiology or medicine) and to Kendrew and Perutz (for chemistry). From now on, the great adventures in molecular biology would *start* from the established facts of the structure of the double helix, the nature of the genetic code and the way it is used in the manufacture of protein, and go on to probe more subtle mysteries of life. So before we take a look at two of the areas of greatest excitement in modern molecular biology this seems to be the place to sum up the understanding of these mechanisms that became established in the 1960s.

# Quantum physics and life

So, life depends on the workings of one particular type of molecule, the double helix of DNA, and the modern

## TABLE 8.1
## THE GENETIC CODE

| first position | second position | | | | third position |
|---|---|---|---|---|---|
| | U | C | A | G | |
| U | Phe | Ser | Tyr | Cys | U |
| | Phe | Ser | Tyr | Cys | C |
| | Leu | Ser | Stop | Stop | A |
| | Leu | Ser | Stop | Trp | G |
| C | Leu | Pro | His | Arg | U |
| | Leu | Pro | His | Arg | C |
| | Leu | Pro | Gin | Arg | A |
| | Leu | Pro | Gin | Arg | G |
| A | Ile | Thr | Asn | Ser | U |
| | Ile | Thr | Asn | Ser | C |
| | Ile | Thr | Lys | Arg | A |
| | Met/ start | Thr | Lys | Arg | G |
| G | Val | Ala | Asp | Gly | U |
| | Val | Ala | Asp | Gly | C |
| | Val | Ala | Glu | Gly | A |
| | Val | Ala | Glu | Gly | G |

To read the genetic code in terms of amino acids the bases have to be taken in triplets. This table enables you to translate any such codon into its appropriate amino acid – for example, the triplet AGU translates as Ser. Three different triplets translate as 'stop'; only one unique instruction indicates the start of a genetic message.

description of living processes at work begins with the structure of DNA. The *origin* of the first living molecules of DNA is, of course, another story. But life today depends, it is clear, on the ordinary behaviour of atoms and molecules in accordance with the same laws of chemistry that govern non-living things. Those laws are rooted in quantum physics. The sugar rings in the backbone of each strand of DNA or RNA, and the various bonds which attach different atoms and groups to those rings, are built up in accordance with the well-known rules of covalent bonding; the two strands of the helix are held together by the quantum effects which give rise to the attraction we call hydrogen bonding. And resonance – a phenomenon purely of the quantum behaviour of atoms – is essential in understanding the basic structures of the molecules of life. All of the processes of life operating inside the cell can be understood as the interaction of complex chemical substances in obedience to the laws of quantum physics.

Replication of the molecule of life itself is one of the easier of these processes to understand, but is still far from trivial. When the double helix unwinds, in response to the action of a particular enzyme which encourages the breaking of the relatively weak hydrogen bonds between the two strands, each strand acts as a template for the construction of a new strand, making two new double helices. There are other enzymes which encourage the nucleotide bases of the growing DNA strands to stick together, the DNA polymerases, and their job is far from straightforward. The carbon atoms around the sugar ring (ribose or deoxyribose) are numbered, remember, in accordance with a chemical convention which gives the label '3' to the carbon in the ring that joins to the phosphate group linking two rings together, and the label '5' to the carbon atom which is not in the ring itself, but on a branch attached to the fourth carbon atom in the ring. This carbon number 5 attaches to the phosphate group next up the chain in the opposite direction. At each end of the strand, of course, there is no phosphate group, and one of these carbon atoms is left attached only to its parent molecule. So there is a difference between the two ends of a DNA strand – one terminates in a 5 carbon and the other terminates, in effect, in a 3 carbon. The ends are labelled the 3 end and the 5 end, and in a double helix of DNA the two strands run in opposite directions, with the 5 end of one strand and the 3 end of the other matched at one end of the double helix.

This might be of no more than passing interest, but careful experiments have shown that when a new DNA strand grows it is always synthesised starting from the 5 end and working towards the 3 end. Since an unravelling DNA helix leaves free one strand starting with a 3 carbon and one starting with a 5 carbon, how can both of them replicate at the same time? In a series of beautiful experiments, Reiji Okazaki showed that only the strand that ends in a 3 carbon is copied smoothly from the free end, so that the *copy* starts with a 5 carbon and polymerises in the 3 direction. The other strand is copied a little more slowly, as a series of fragments each of which is copied in the correct $5 \rightarrow 3$ direction, in effect 'backwards' from the unwinding helix, starting at the fork where the helix is unwinding and proceeding a little way towards the free end of the strand. Then, as more helix unwinds another fragment is copied, and the separate fragments are subsequently joined together by another enzyme to produce a second continuous strand of newly synthesised DNA.

I have gone into the details of this one example to highlight the way in which even the subtle complexities of life in the cell fit within the framework of quantum physics and chemistry. Even this is a simplification of the story, and I cannot go into even such simplified detail for most of the other mechanisms that ensure not only the faithful copying of DNA but also the faithful translation of DNA code into protein. For example, one fascinating discovery is that a short fragment of RNA is synthesised first when DNA replicates, and that the new DNA strand actually grows from the 3 end of this RNA primer, which is itself discarded when the DNA strand is complete. And the action of unwinding enzymes themselves is now well understood in terms of reactions involving ATP. For the details, you'll have to turn to a modern biochemistry text.* But I can summarise briefly the steps in the process by which proteins are manufactured, bearing in mind that each step along the way involves detailed chemical processes that are now beginning to be properly understood.

The first stage is the synthesis of a strand of messenger RNA, a single RNA molecule that is an exact counterpart to part of one strand of the DNA in the double helix of one particular chromosome. Somehow, the DNA untwists in just the right place, and by just the right amount, to allow the manufacture of an mRNA copy of a stretch of genetic code – a gene – providing the information needed to make one protein. The chemical reactions require energy, which is provided from high-energy phosphate bonds. Each nucleotide base which is used in the synthesis of the mRNA comes attached to a double phosphate group, a pyrophosphate, and is called an activated nucleotide (ATP is the activated form of the adenine nucleotide). Once synthesised, the mRNA moves out of the nucleus into the cytoplasm of the cell, and the chromosomal DNA coils back up, storing the fundamental information until it is needed again. The bases strung along the sugar-phosphate backbone of mRNA include uracil in place of the thymine found in DNA, but even this difference is accommodated within the rules of quantum physics, because the power of uracil to form hydrogen bonds is exactly the same as that of thymine, and it is the hydrogen bonding that is important in passing on the information from the gene. Those mRNA hydrogen bonds will soon be used to pair up with other bases and thereby direct the synthesis of a growing polypeptide chain.

Those other bases, individual triplets that each make up one word in the genetic code, a codon, are each attached to one end of one molecule of another kind of RNA, transfer RNA. The other end of each tRNA molecule latches on to a molecule of one of the amino acids used in the manufacture of a protein, the amino acid that corresponds, in the

---

* Such as Lubert Stryer's *Biochemistry*, W. H. Freeman, San Francisco, second edition, 1981.

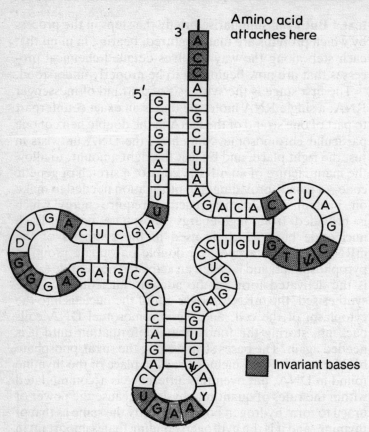

Figure 8.5 The structure of a molecule of transfer RNA. Such molecules always contain about 80 bases linked in a single chain, and a specific amino acid attaches to one end, which always ends with the sequence CCA. As well as the regular bases found in messenger RNA, tRNA contains unusual components indicated here by symbols other than G, U, A, or C. Hydrogen bonds hold the tRNA molecule into a particular shape, with a triplet of unpaired nucleotides, called an anticodon, at the opposite end to the amino acid. This anticodon is used to plug the tRNA in to the appropriate mRNA codon. (Reproduced, with permission, from Figure 15.6 of Curtis, *Biology*, op. cit.)

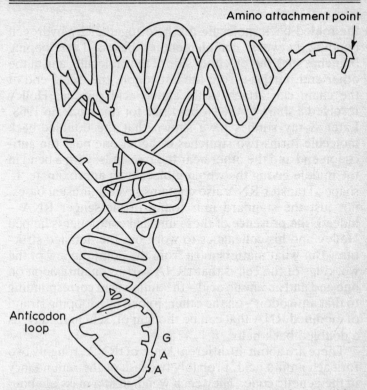

Figure 8.6 The actual shape of a tRNA molecule is something like this.

genetic code, to the codon carried by that particular tRNA. Or rather, strictly speaking, to the *anti*codon. The triplet of bases on the tRNA molecule is the complementary mirror image of a set of bases on the messenger RNA, and it is a codon in the mRNA which matches up to an anticodon on the tRNA. In 1965, a team headed by Robert Holley, at Cornell University, determined the complete structure of a transfer RNA molecule – as it happens, one from yeast, although others have since been worked out in similar detail and shown to have the same general structure. They showed that this structure is basically a single strand of RNA which loops back on itself, in a rather sloppy fashion, so that there is a hairpin bend on one end, where the anticodon triplet is prominently displayed, and stretches of the folded back molecule are held together by hydrogen bonding between complementary bases. Two looping branches stick out on either side of the molecule, and at the other end, farthest from the anticodon, the free 3 end of the chain can attach to an amino acid residue. Holley received a share in the Nobel Prize for this work, in 1968. Later X-ray studies have shown that the doubled-back molecule forms two stretches of helix, one near the anticodon end and the other near the free ends, with a bend in the middle giving the whole molecule an approximate 'L' shape. Transfer RNA also contains several unusual bases, not just the standard four found in messenger RNA – indeed, the presence of these unusual components helped Holley and his colleagues to work out the detailed structure. But what matters most from the point of view of the workings of the cell is that tRNA carries an anticodon on one end and an amino acid – the amino acid corresponding to that anticodon – on the other, joined by a looping strand of modified RNA that can be thought of, schematically, as a doubled-back helix.

There are about 40 different kinds of tRNA, roughly two for each amino acid, probably because of the redundancy in the genetic code. They each weigh in at a mass of about 25 000 daltons, and they each latch on to their own amino acid by means of a high-energy covalent bond, whose energy comes from a molecule of GTP, guanine triphosphate, which is similar to ATP. They are fascinating molecules because of their unusual structure, because they provide, literally, the link between nucleic acids and amino acids, and because their unusual role in the cell sees them behaving like enzymes – carrying out a specific function – even though they have the fundamental structure of a nucleic acid, an 'instructing' molecule. It may be that tRNAs themselves were among the first molecules of life to appear, and that it is their properties which have in large measure defined the progress of molecular evolution – in one direction to produce more efficient information storing molecules, DNA, and in the other to produce more efficient worker molecules, proteins – since the earliest appearance of life on Earth. But protein synthesis in the cell today requires more than just information, from mRNA, and a source of the right amino acids, from tRNA. It needs a building site where the information can be used to construct long polypeptide chains, and that building site is the ribosome.

Ribosomes are really big molecules. They mass about 3 *million* daltons, and more than 60 per cent of this mass is RNA, ribosomal RNA. Both rRNA and tRNA are themselves, of course, manufactured by the cell using information stored in its DNA. A long chain of RNA made from the DNA template may be chopped up into different pieces (cleaved) by appropriate enzymes, and then rearranged (with an addition of about one-third protein) to provide one ribosome and one tRNA molecule. Each ribosome comes in two parts, one containing an RNA molecule massing about a million daltons, the other half as big, joined together to make a lopsided sphere with a groove – the 'equator' – running not around its middle but one-third of the way in from one side of the structure. Molecular biologists are beginning to determine details of just how the ribosome uses the information in mRNA and the amino acids carried by tRNA to make polypeptide chains, but I won't go into those details here. The simple picture is something like this.

A ribosome latches on to a molecule of mRNA, which may be thousands of bases long, at the 5 end and 'reads' the first three bases of the genetic message coded by that molecule. From the soup of active chemicals in the cytoplasm, it then selects a molecule of tRNA which has the appropriate anticodon to match up with this three letter word, and temporarily holds this in place, with the aid of hydrogen bonds, on the mRNA molecule. This is only possible because of the way the triplet of bases on the codon on the mRNA match up and form hydrogen bonds with their counterpart bases on the anticodon. At the other end of the tRNA molecule held by the ribosome and the mRNA, there will be an amino acid, the one specified by the triplet of bases on the mRNA. This amino acid is attached to the tRNA by a high-energy bond. Now, the ribosome 'reads' the next triplet of bases, the next word in the genetic message, and lines up the appropriate tRNA, and its amino acid load, alongside the first. The two amino acids brought alongside one another in this way are then detached from their tRNA molecules and joined to each other by enzymes, making use of the energy available from the high-energy bonds. The first tRNA molecule is released to seek out another amino acid partner in the cytoplasm, and the ribosome moves along the messenger RNA to read the next triplet of bases and repeat the whole process. Gradually, a polypeptide chain is built up (always from the amino end to the carboxyl end). As the ribosome moves down the molecule of messenger RNA, reading its message and translating it into polypeptide terms, another ribosome begins at the 5 end and repeats the whole process. Several ribosomes may be at work on one molecule of mRNA simultaneously, and hundreds of identical protein molecules may be manufactured from each molecule of mRNA.* As each ribosome gets to the end of the message, it releases a completed polypeptide chain which folds up, with the aid of other enzymes, into the specific three-dimensional shape that gives that protein its special properties. The ribosome is then free to seek out another molecule of mRNA – *any* molecule of mRNA – and manufacture another protein – *any* protein. At every stage, the process depends directly upon the rules of quantum chemistry. Hydrogen bonds are made and broken; energy is provided to make strong covalent bonds; even the final folding of the protein molecules places each of them in a

---

* And hereby hangs a tale that I can't resist going into in a little more detail. How does the cell avoid getting swamped with one particular protein, as ribosomes diligently follow one another along a molecule of mRNA, busily translating its message *ad infinitum*? It seems that the 3 end of each fresh molecule of messenger RNA has attached to it a string of molecules – a tail – that has nothing to do with the genetic code. As translation proceeds, this tail gets shorter; each ribosome chops off a fragment as it reaches the end of one translation of this particular gene into protein. When the tail has completely gone, after hundreds of ribosomal passes along its length, enzymes break up the molecule of mRNA into its constituent parts, which are recycled by the cell. No more of that particular protein is manufactured for the time being.

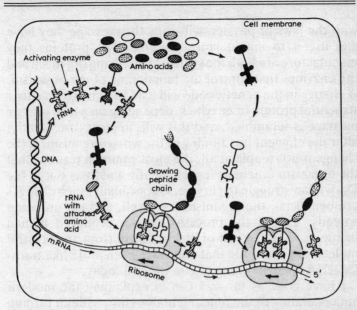

Figure 8.7 Protein synthesis in a bacterial cell. DNA provides the instructions for the manufacture of at least 32 different kinds of tRNA. Each of these latches on to a specific amino acid and carries it to the site where a ribosome is reading mRNA, itself transcribed from the DNA. There, the amino acids are joined in the right order to form the proteins the cell needs. The ribosome moves along the mRNA joining amino acids on to the chain in sequence until the polypeptide is complete. Then, it is released to do its work. (Based, with permission, on Figure 15–20 of Curtis, *Biology*, *op. cit.*)

configuration that is the most stable, in energy terms, for that molecule. And, at last, we can see how quantum physics is responsible for the changes in the genetic code – mutations – that are the driving force of evolution.

One letter in the genetic code of a DNA molecule may get changed by an accident – a photon of ultraviolet light may damage the molecule, a base may be copied incorrectly during cell division, an extra base may be inserted or one deleted in error, and the order of bases may get shuffled up during recombination. Whatever happens, the mechanisms of the cell, operating blindly in accordance

with the laws of physics, will copy the message they have
and use it to manufacture proteins. The proteins they
manufacture are both the structure of a living organism and
the enzymes that control the functioning of the organism.
A change in the genetic code will lead to a change in either
structural protein or enzymes, depending on which genetic
message is scrambled, and this will, in most cases subtly,
alter the efficient functioning of the whole organism. If the
change is noticeable at all, the most common result is that
the organism functions less efficiently and loses out in the
Darwinian struggle to survive. Occasionally, such changes
are beneficial, the organism does well, and the mutation
spreads. This is the process that has produced human
beings, and all the rest of life on Earth, from an ancestral
molecule, a molecule that may have been more like transfer RNA than anything else in the cell today.

I have gone as far as I can in explaining the modern
understanding of the fundamentals of life, without turning
this book into another biochemistry text. The time has
come to move out again from the inner workings of the cell
and to see how these subtle, and sometimes not so subtle,
changes at the molecular level really do affect whole organisms, including the human organism.

# CHAPTER NINE

# JUMPING GENES

Most of the work which led directly to the Crick and Watson model of DNA as a double helix, and the bio-chemical attack which cracked the code of life, used biological material from the cells of prokaryotes. These are simple, single celled organisms – such as bacteria – which have their DNA, usually a single strand, floating around loose inside the cell. The simplicity of the structure and of the DNA itself made it easy (relatively speaking) to extract the DNA and to investigate its properties. But virtually all of the forms of life we can detect with our human senses, including ourselves, are made in a rather different way. It isn't just that these larger life forms are aggregates of cells working together; the cells themselves are very different from prokaryotic cells. Our cells, and those of other multi-cellular organisms, have their DNA packaged into a cell within the cell, a nucleus, which gives this type of cell its name, eukaryote, from the Greek for the kernel of a nut. The reasonable assumption is that prokaryotic cells came first in the course of the evolution of life on Earth, and this assumption is reflected in their name.

The eukaryotic cells of multicellular organisms contain a lot more DNA than the prokaryotic cells of bacteria. A bacterium needs enough genetic information to control the chemical factories of one cell, and to ensure accurate copying of one strand of DNA and separation of the two resulting strands into two daughter cells when the cell divides.

That is all. But every cell in your body contains a full genetic blueprint describing both the development of the complex structure of the whole body from a single fertilised egg and the operation of all of the specialised cells in that structure to make the adult form function efficiently. In any particular cell, far more than 90 per cent of that information is never used. It stays locked up in DNA strands wound tightly into chromosomes inside the nucleus of the cell. Just the small amount of genetic information relevant to the running of that particular cell – how to be a liver cell, or a muscle cell, or whatever it may be – ever gets translated into RNA which in turn is used to control the manufacture of proteins. But every cell still carries a complete set of chromosomes, which is why, in principle, it would be possible to clone a human being, producing an exact copy from a single cell taken from the human body and triggered (somehow) into developing in the way a fertilised egg develops.

But that 'somehow' glides over a host of complex issues. From that fertilised egg, after a very few simple divisions of the cell and copying of DNA, different cells begin to develop differently. Even as the embryo's cells are dividing and growing at a rapid rate, they are specialising, preparing for their role in the body of the baby that will be born a few months later, a role they will play throughout its life. Something – it can only be part of the genetic message itself, information obtained from the DNA on the chromosomes and used in response to the immediate environment of each cell – 'tells' each cell how it must develop, to become muscle, liver, brain or whatever. This subtle biochemical control of the development process is far from being understood. But progress is beginning to be made in understanding the later part of the story, how, after the cells have differentiated and developed into the mature form, they stay differentiated, so that your liver, for example, doesn't suddenly decide that it ought to develop cells characteristic of, say, brain tissue. Clearly the whole business depends on the way parts of the genetic message coded in the chromosomes, and only the right parts, are translated into RNA and then proteins. The puzzle is to find out how parts of the genetic code are switched on and off. And as well as its interest in terms of pure biological understanding of what we are and how we came to be what we are, this line of research holds out the obvious possibility of first understanding and then, perhaps, controlling the disease known as cancer, a disease in which cells stop obeying the part of the genetic blueprint appropriate for their role as part of a multicellular organism, and do start dividing and spreading unchecked, reverting to something like the ancestral behaviour of their prokaryotic forebears.

# Packing in the DNA

Some idea of the efficiency and complexity of the natural processes which read DNA off chromosomes can be gleaned from the understanding of chromosome structure which has been developed in recent years. Without going into details of how the structure has been determined, it looks something like this.*

DNA and proteins are combined in the material structure of a chromosome, called chromatin. Remember that fifty years or more ago most biologists thought that DNA provided a scaffolding for protein molecules that carried the genetic information. They had the story upside down, for in fact this particular family of proteins, called histones, provides the scaffolding on which the DNA is tightly wound and packed efficiently into a very small space. A cluster of eight histone molecules forms a bead-like structure round which the DNA double helix as a whole makes two loops, like a rope being wrapped around a basket ball. Another histone fits over the two loops of DNA, clamping

---

* The story is told in more detail in modern texts such as Lehninger, but some of it is from recent research and hasn't yet filtered through to all the textbooks, let alone popularisations. I have followed the account given by Lubert Stryer, *Biochemistry*, Freeman, San Francisco, Second Edition, 1981.

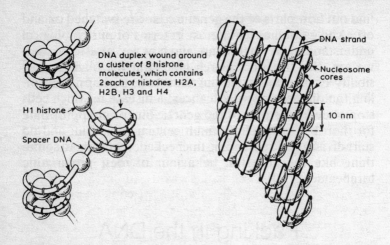

Figure 9.1 DNA double helices are wound around histones to make a structure like a bead necklace, which is then wound around itself to pack the genetic material still more compactly in the chromosomes. A nanometre is one billionth ($10^{-9}$) of a metre. (Adapted, with permission, from figures in *Modern Genetics*, by F. J. Ayala and J. A. Kiger, Jr, Benjamin/Cummings, Menlo Park, California, 1980, 1984.)

them in place on the bead, and on either side of the bead there is a short stretch of spacer DNA bound to another histone and linking the DNA-wrapped bead to similar DNA-wrapped beads on either side. Each of these 'beads' is called a nucleosome, and because they are joined to one another by a flexible DNA-histone link the whole collection of nucleosomes associated with a length of DNA can be coiled up compactly, very much like the coiling of a bead necklace, to make an even more compact structure, a variation on the helical theme, which can itself be coiled (or supercoiled) still further. This is a masterpiece of packing. Each cell in your body contains enough DNA to stretch over a distance of 180 cm – comparable to, or probably longer than, your height. Except when it is unwound for copying as a cell divides, this DNA is packed into 46 tiny cylinders which, laid end to end, would still only have a total length of 0.2 millimetres. In round terms, the DNA is packed into a length one ten-thousandth of its 'natural' length. Somehow, in each cell, just the parts of each chromosome that are relevant (not necessarily every chromosome) are slightly unwound as and when the information they carry is required, then the beads are packed tightly away into the body of the chromosome once again. Without this careful packing, so much DNA floating about in the cell would provide an impossible housekeeping problem – it would get broken, and the machinery of the cell would need more luck than judgement in finding the right bits to use. But, perhaps the most remarkable achievement of the cell is the way it can uncoil *all* of this material, copy it faithfully (starting the copying process at many sites simultaneously on each long strand of DNA), divide, taking one copy of the DNA into each daughter cell, and pack up the genetic material back into tightly wound chromatids, all in the space of a few minutes and with only the very occasional copying error. This process is going on all the time in your body.

But we are not going to be able to work out how the cell does all these tricks by looking in closer and closer detail at the structure of the DNA and the way it is wound up into chromosomes. We have to stand back from the detail and look at the broader picture, at chromosomes and stretches of chromosome (genes) as the basic units of heredity and mutation, and at the effect on the whole organism – the phenotype – of changes in the genotype. In a way, this is a return to the 'old fashioned' kind of genetics that made great progress in the decades before the Second World War, and before the revolution in molecular biology. That is one reason why some of the key developments in this area were not fully recognised or appreciated at the time they were first made, in the 1940s. Today, however, the value of the study of the whole organism when dealing with evolution and mutation of eukaryotic organisms, such as ourselves, is clear. That is why the pioneer who made the unsung breakthrough in the 1940s, Barbara McClintock, at last received the Nobel Prize for her work in 1983. Both approaches, the molecular and the organismal, gain from each other. And now that we have gleaned some idea of what is going on at the molecular level, we are well equipped to understand, more easily than her colleagues could in the 1940s, the importance of McClintock's investigation of certain mutations in her maize plants.

# McClintock's maize again

Barbara McClintock didn't have to know whether it was protein or DNA that carried the code of life on the chromosomes of her plants. All she had to know was that chromosomes do indeed carry the genetic information, and that each chromosome is made up of many different genes. Rather than thinking of those genes as long stretches of DNA double helix, we can think of them, once again, as the shortest functional stretches of chromosome. Working at Cold Spring Harbor in the early 1940s, McClintock was studying changes in the genotype – mutations – which produced changes in the pattern of pigmentation, the colouring, of the leaves and kernels of the growing maize plants. The discovery that opened the door to her new insight into how genes are controlled – how DNA, as we now know, is switched on and off – came from a simple observation, that in one family of plants there were patches of colour that didn't belong in the leaves.

In most plants, the leaves are one colour, usually green or, in some strains, pale yellow, light green or even white. In this unusual strain, however, there were splashes of different colour, a streak of normal green on a light green leaf, a spot of yellow on a normal leaf, and so on. The interesting thing about this is that each leaf has developed from a single cell at the stem of the plant, the leaf growing due to the repeated division and multiplication, with appropriate differentiation, of this cell and its daughters. The spots of misplaced colour could be traced to an origin in one of these daughters, a cell which suffered a mutation and then reproduced its own daughters complete with that very slightly different set of genetic instructions, growing into a patch of the wrong colour among the sea of 'normal' (for that leaf) cells. Such mutations provided McClintock with a tracer that could identify exactly when, in which cell and at what time during the process of development and differentiation which made a leaf an integrated whole, a particular mutation had occurred. She also found that in some cases these multicoloured leaves showed a different mutational behaviour from the plant as a whole, either faster or slower rates of mutation. Presumably, this change in the frequency of mutations was itself a mutation, and derived in the same way from a single dividing cell early in the process of leaf differentiation. Similar effects are seen in the corn cobs of the maize plant, spots of colour among the yellow kernels that, thanks to the efforts of farmers and the food marketing industry, we think of as normal when we buy corn in the supermarket.

After several years of painstaking study, similar in many ways to the work of Mendel, who also worked with plants and had to wait for a year between generations in order to trace the patterns of mutation, by 1947 McClintock was convinced that she was watching two different, but related, control genes at work among the chromosomes of the maize plants. The genes which produce the observed structural characteristics of a plant – or a person – must be controlled by other genes, as we have seen, because only a few genes in each cell are used at any one time. McClintock explained her observations in terms of two kinds of control genes. One was a gene which sat next to the structural gene (in this case, the gene for colour) on a particular chromosome, and turned the structural gene on and off. A second controlling element, as McClintock called these genes, was responsible for determining the *rate* at which the first control gene operated, speeding up or slowing down the frequency with which the switch controlling the structural gene was turned on and off. And painstaking studies of the effect of recombination on the pattern of behaviour, as well as studies of the chromosomes themselves under the microscope, showed beyond doubt that while the first gene – the switch – sat next to the structural gene it controlled, the second control gene – the regulator – could be located almost anywhere, far away on the same chromosome, or even on a different chromosome entirely. The continuation of this work in the years up to the end of that decade, the 1940s, convinced her that these controlling elements were not even fixed in a unique location on the chromosome, but could jump from place to place and chromosome to chromosome. The control genes could move to different places, bringing different structural genes under their control and affecting the subsequent development of all of the offspring of the cell they inhabited. Such discoveries, with the benefit of hindsight, seem to us today, as they did to McClintock in the 1940s, clearly to point the way to a better understanding of the whole business of development and differentiation, as well as addressing the problem of cancer. But when she presented her results publicly for the first time at the Cold Spring Harbor Symposium in the summer of 1951, she met with a blank response.

There were many reasons. Many of the people she was talking to were a new breed of biologists, former physicists and chemists who were excited about molecules of life, and lacked the basic training or patience to cope with the complexities of understanding the genetic behaviour of a multicellular eukaryotic organism. McClintock had lived with her plants and for her work for decades; she understood them and had a feeling for the organism unmatched even among biologists of the old school, but they were incomprehensible to the new breed. Simple bacteria, phage and experiments that could be done in the test tube, with cultures that reproduced every few minutes, not once a year, were not only fashionable but proving invaluable in unravelling the secrets of life. Complex eukaryotic organisms were unfashionable, but for good reasons. If it took 20 years to get new insights from studying maize, and two years to make a major discovery using phage, then as long as new discoveries were clearly there to be made using phage bright young biologists (and physicists and chemists) would continue to move in that direction.

The idea that genes could jump about from place to place probably seemed ludicrous to any of this breed of molecular biologist who thought about it at all, while it was also more than a little esoteric and abstract for anyone used to working with living cells that only had one strand of DNA and a few genes anyway. The complexities

McClintock was studying literally didn't exist in the forms of life most people involved in the search for the double helix were studying. And even if her evidence were taken at face value, she could offer no detailed, biochemical explanation of how the controlling elements worked, how structural genes were switched on and off. That could only come from an understanding of the molecules themselves, and we can imagine our hypothetical molecular biologist in 1951 thinking that the time to worry about control genes and jumping genes would be when the molecules were understood well enough to identify the chemical switching process. Probably nobody did think it through in this way at the time, but in truth that is how things did develop. Like Mendel, McClintock was ahead of her time, and her work had to wait for recognition until quite different attacks on the puzzle of the nature of life made progress. Only then could it be seen in its proper context. Unlike Mendel, however, McClintock lived to see her work take its proper place in science.

# The French connection

Throughout the 1950s, and indeed right up to the present day, McClintock continued her work, developing it further, reporting regularly to the Cold Spring Harbor Symposia, but scarcely bothering to publish elsewhere, and (in the 1950s at least) being largely ignored. But although the problems of the control of genes and the choice of which genes are in use – expressed – at any one time are far less in simpler bacteria than they are in complex eukaryotes like us, they do still arise. And, also during the 1950s, Jacob and Monod, with the team at the Pasteur Institute in Paris, were puzzling over the way *E. coli* bacteria made the most efficient use of the food resources available to them.

*E. coli* can obtain the carbon atoms they need to maintain the chemistry of life from a variety of sources, but one food which can supply all of their carbon needs is the sugar lactose. Among the genetic material of an *E. coli* cell, on its single 'chromosome', there is a sequence of three genes that code for three proteins. One of these is an enzyme called beta-galactosidase, whose job is to chop up lactose (the sugar found in milk) into its two components, galactose and glucose, an essential step in the process of breaking down the food into usable pieces.

When *E. coli* finds itself in an environment where there is a supply of lactose, it needs beta-galactosidase, and plenty of it, in order to make the most of the opportunity. So this gene system must have the potential for rapid production of large quantities of enzyme. But if there is no lactose to be digested, then manufacturing this particular enzyme is worse than useless, because it uses up the resources and energy of the cell without providing any benefit. Sure enough, evolution has selected bacteria that only make the enzyme when they need it. In an individual cell of *E. coli* living in a culture medium where there is no lactose there are less than ten molecules of beta-galactosidase. But if the same bacterium is transferred to a growth medium containing lactose, then it proliferates and each of the daughter cells in the growing culture is found to carry several thousand copies of the enzyme. Somehow, the presence of the food triggers the manufacture of the enzyme needed to digest that food – a perfect example in miniature of gene control at work.

One of the other two genes in the sequence of three is also involved in lactose digestion, producing an enzyme that concentrates lactose inside the cell, passing it through the membrane that surrounds the cell; just what job the enzyme coded for by the third gene does we don't know, but it seems likely that it too is involved in making use of lactose, since the French work showed that the three genes work as a group. When more beta-galactosidase is produced, more of the other two enzymes is also produced, in exact proportion. Halve the production of beta-galactosidase and you halve the production of each of the other two enzymes as well. Jacob and Monod deduced that all three genes must be under the control of a fourth gene, sitting alongside the group on the chromosome. This they called the operator. Under normal conditions, the argument ran, the operator prevented the translation of the three genes alongside it into protein. But the presence of lactose activated the system. How?

From studies of mutant strains of bacteria with a different response to the presence of lactose, Jacob and Monod were led to exactly the conclusion McClintock had reached a decade previously during her studies of maize. There must be yet another gene, somewhere else, which was responsible for regulating the activity of the operator. They called this the regulator, and gave the name 'operon' to the whole system of regulator, operator and the set of structural genes controlled by the operator.

All of this was predicted, without the regulator or operator being identified chemically, and published in Jacob and Monod's classic paper of 1961.* The fact that that paper makes no mention of McClintock's work, reported exactly ten years earlier, indicates neither unusual ignorance on the part of the French pair nor any scurrilous attempt to claim credit for an idea that had already been aired. In 1961, the *general* ignorance of McClintock's work among molecular biologists made it far from natural to make the connection immediately. But when it was pointed out the connection was quickly acknowledged, and this marked the beginning of the process which brought McClintock's work in out of the cold.

It was still a slow process, for at first the operon system was just a beautiful theoretical construction with no hard chemical evidence to back it up. It was only in 1966 that Walter Gilbert and Benno Müller-Hill, working at Harvard, were able to identify the repressor molecule itself, a protein that binds to the chromosome at the start of the stretch of DNA coding for the three enzymes. The present understanding of how the system works includes the up to date understanding of the role of messenger RNA, itself predicted on theoretical grounds by Jacob and Monod, but not identified chemically, in 1961. At the start of a sequence of DNA coding for a gene or a set of genes there is a particular short stretch, called a promoter, which is

---

* *Journal of Molecular Biology*, volume 3, page 318, 1961.

recognised by the enzyme that manufactures messenger RNA (RNA polymerase) as the start of the message. RNA polymerase can only attach to the DNA and begin to manufacture mRNA at such a site. In the *E. coli* system we are interested in (called the lac system, for short), the operator postulated by Jacob and Monod sits between the promoter and the beginning of the sequence of DNA that codes for the three enzymes. The operator is not so much an active gene, after all, but another site to which a particular type of molecule can become attached. That molecule is the large protein identified by Gilbert and Müller-Hill. It is manufactured in the cell in accordance with the instructions coded for by the repressor gene, which doesn't have to be anywhere near the scene of the action, because the protein molecules it produces have an affinity for the operator site and eagerly bind to it. (Nor, of course, does this repressor gene have to produce very much protein, so it isn't wasteful of the cell's resources.) RNA polymerase can still attach to the promoter site, but when it tries to move on to read the DNA code in the genes alongside and manufacture RNA, its movement is blocked by a large protein molecule attached to the operator site.

So far, so good. But what happens when there is lactose about and the cell needs those three genes working flat out? The protein molecule that binds to the operator site may find that site attractive, but it is even more attracted, in chemical terms, to lactose molecules. When there is lactose about, the protein releases its grip on the chromosome and latches on to a lactose molecule instead. Any other protein molecules produced by the repressor (and there are only a few of them) do the same. The RNA polymerase has an unobstructed path, and can proceed efficiently about its task of reading DNA and manufacturing mRNA which itself is used to make the enzymes the cell needs. When all the lactose has gone, a protein molecule with nowhere more attractive to go re-attaches to the operator site, and production of these enzymes ceases.*

---

* You may think this tale, interesting though it is, wanders a little far from my theme of quantum physics and life. But why does a protein

By the late 1960s, it was clear that even in simple prokaryotes structural genes are controlled by other genes to
ensure the efficient functioning of the cell. But still the idea
of genes that could jump around from one chromosome to
another was anathema. It was only in the next decade, the
1970s, that molecular biologists began to come to grips
with the complexity of the control and regulatory processes, following the discovery of genes that jump around
in bacteria (and thereby making McClintock's work with
maize suddenly fashionable, respected as a true work of
genius three decades ahead of its time) and, an equally
shocking heresy, finding large stretches of DNA that don't
seem to code for anything useful at all, but occur so pro-
lifically in the genome of many species that the code for a
useful, functioning gene may be split into parts separated
by chunks of 'nonsense' DNA, chunks of gibberish which
have to be excised from the mRNA copied from the DNA
before the mRNA itself is used as a guide to the manufacture of protein. The enormous complexity of the activity
going on at this level inside the cell can only be described as
a separate evolutionary system, going on at the molecular
level today as it has since life first appeared, and to a large
extent (in terms of DNA) totally indifferent to what goes
on outside the cell. It is a sobering thing to discover that a
great deal of the DNA in your own body cells, and the
copying and replication activity it is involved in, doesn't
have anything to do with the smooth running of your body

---

molecule find one site more attractive than another to bind to? Because
of the interplay of forces, at the molecular level, and its 'need' to seek
the lowest energy configuration available. And what are the rules determining how those chemical forces interact, and which state is the lowest
possible energy state? The laws of quantum physics. Although I am not
spelling out the details of the quantum physical behaviour of each and
every atom and molecule at every stage of the discussion, the quantum
rules are the only rules that those atoms and molecules live by. Seek to
distribute your electron cloud in a configuration with the lowest possible
energy, and you will achieve molecular nirvana. All the complexity of
life, and the proud achievements of the human intellect, depend on this
ultimate cosmic truth.

after all, but only with the selfish manipulation of the cell mechanisms by the DNA to ensure its own survival. It is a story which is still being deciphered, and which will make a wonderful book in about ten years' time. Meanwhile, here is a brief glimpse at the wonderland molecular biologists are beginning to explore inside eukaryotic cells.

## McClintock rediscovered

The rediscovery of jumping genes didn't come all at once. There were experiments with *Drosophila* which showed mutations that are now recognised as due to changes in the genome such as McClintock was describing in maize, but nobody seems to have made the connection at the time. The new breed of molecular biologists weren't ready to acknowledge such striking new ideas until they appeared in the simple organisms, bacteria and phage, with which they were by now so familiar. And appear this pattern of genetic behaviour certainly did. Even in the early 1960s, some experiments with phage had shown that the virus invading a bacterial cell could insert its genetic message almost anywhere in the bacterial chromosome, and that when the message was used to make more phage they would in turn infect other bacteria, splicing themselves in more or less at random to the newly invaded bacterium's stretch of chromosomal DNA.* So there must exist in the cell the potential to manufacture enzymes that could snip apart the strands of DNA and splice in new chunks in almost any place. A few years later, attention focused on a different kind of change going on in the DNA of *E. coli*. These mutations affected several functions of the *E. coli* at once, a change at one site on the chromosome – one locus – apparently influencing other genes nearby. The discovery led to unambiguous identification (and, equally important,

* First reported by A. L. Taylor, *Proceedings of the National Academy of Sciences*, volume 50, page 1043, 1963.

recognition) of mobile control genes at work. The mutations were recognised as the insertion of bits of genetic material from somewhere else on the chromosome into a new site, regulating the activity of genes that came under the influence of the inserted gene. The discovery moved on by one step from Jacob and Monod's static operon system, and it was a crucial step.

The pace of research quickened in the 1970s, with the development of recombinant DNA techniques, the techniques of genetic engineering. By the late 1970s, researchers were able to analyse the sequence of bases on lengthy stretches of DNA as a matter of routine, identifying exactly what the genetic code said; they could manufacture pieces of artificial DNA dozens of bases long (by 1982, they could synthesise a gene for interferon, 514 pairs of bases in a precise order); they had identified and could use the enzymes that cut DNA; and they had also identified and used the enzymes that stick cut ends of DNA back together, with the artificially created 'gene' inserted into the gap. The astonishing potential these developments have created for human interference in the genome of living organisms, including the human genome, has been the subject of intense debate in the 1980s. On the one hand, it holds the prospect of incalculable benefits, including new drugs such as interferon, new sources of insulin for diabetes sufferers, new strains of food crops (and animals) and perhaps the correction of inherited genetic defects, such as sickle-cell disease and hemophilia. On the other, some people have argued, it raises the possibility of Frankenstein monsters, and of mutated genes escaping from the lab to cause new plagues. The genetic engineering debate is outside the scope of this book.* The point I wish to make is that, first, the development of all these techniques enabled

---

* The best guide to recombinant DNA techniques available so far is *Recombinant DNA*, by James Watson, John Tooze and David Kurtz (Scientific American Books, 1983). A slightly rose-tinted but very clear 'popular' account of the genetic engineering revolution is given by Steve Prentis in *Biotechnology* (Orbis, 1984).

biologists to study the behaviour of genetic material inside the cell with previously undreamed of ease; secondly, it turns out that all of this cutting, splicing and insertion of new genetic material into old chromosomes is not unnatural at all, but simply human mimicry of the natural processes that go on inside the cell all the time.

It was against this background that McClintock's work was first recognised and then honoured. We now know much more about jumping genes; not least being the fact that the gene doesn't really 'jump' at all, because the original stays in place on the chromosome. A copy of the gene is manufactured, using the cell's usual machinery for replicating DNA, and this copy moves to another site where the chromosome is cut open and the gene is inserted. But this transposable element consists of a little more than a naked gene, and has some more tricks up its sleeve. First, each jumping gene is flanked by short stretches of DNA, complementary to one another, so that when the whole stretch of DNA including the gene and these inverted repeats is cut loose the ends pair up in a helix, producing a lollipop-shaped structure in which the twisted ends form the handle and the mobile gene loops across the middle. These genetic lollipops can be seen, with the aid of the electron micrograph. Such a mobile element is called a transposon. And the transposon must carry with it not only the 'active' gene, but genes that work for it by ensuring the manufacture of the enzymes that do all the cutting and splicing when the transposon is inserted at a new site. The cutting and splicing, however, is not always perfect. Occasionally, the copy of the transposon will carry with it a copy of some of the neighbouring DNA from its original site; sometimes, during the cutting and splicing at the insertion sites bits of the 'host' DNA get lost. The jumping genes, as well as their direct influence as control genes on their new neighbours, can thus introduce all of the standard mutations – deletions, insertions, and inversions. Their main role may be to control their neighbours, but as a byproduct of this they speed up the rate of mutation, and thereby hasten the pace of evolution.

At first, all of this activity was thought to be exceptional, aberrant behaviour in a few bugs studied in the lab. But as the evidence mounted it slowly dawned on biologists that in fact it is the norm, and that the more complex eukaryotic cells of higher organisms such as ourselves must undergo far more of this shuffling of the genetic material than bacteria, which have very little DNA to shuffle. McClintock's biographer, Evelyn Fox Keller, identified the moment when the light finally dawned as the summer of 1976, when the Cold Spring Harbor Symposium took as its theme 'DNA Insertion Elements, Plasmids and Episomes', and 'explicit acknowledgement of McClintock was made in introducing the term "transposable elements" to refer to all "DNA segments which can insert into several sites into a genome".'*

By 1981, a meeting on evolution held at King's College in Cambridge came out loud and clear with the message that the genomes of higher organisms are in a state of dynamic change, with considerable re-arrangement of the genes among the chromosomes being the norm on an evolutionary timescale, and perhaps being a driving force of evolution. It had become clear that just as human intervention using recombinant DNA techniques could take artificial genes and insert them into chromosomes, so viruses might transmit genes from one host to another during their own life cycle. This is straightforward enough to understand when a phage that has insinuated itself into a bacterium causes copies of itself to be manufactured and these copies invade other bacteria – a simple error could tag on a bit of bacterial DNA to the copy of the phage DNA. It is far more startling when evidence for this kind of sideways translation shows up in totally different species. At that meeting in Cambridge Alec Jeffreys, of Leicester University, drew attention to a protein called leghemoglobin, which is used by the plants known as legumes during nitrogen fixation. The gene for leghemoglobin looks very much like the gene for globin, an *animal* gene coding for the protein in hemoglobin. Jeffreys suggests that this animal

* *A Feeling for the Organism*, page 187.

gene was translocated into the ancestral form of the plant, relatively recently during evolution, as a passenger on a virus. And that possibility opens up dramatic evolutionary possibilities, even if it only occurs rarely.*

This new wonderland of evolutionary possibilities depends on one crucial ability of the cell, its ability to cut pieces of DNA out of the chromosome and splice them in somewhere else. Why and how should this skill have evolved? Such an ability is absolutely crucially important during sexual reproduction, making possible the genetic shuffling that goes on during recombination. But that is rather begging the question of the origin of this ability, since it must surely have developed first, with the benefits of sex and recombination evolving to take advantage of these mechanisms. Indeed, the mechanisms are probably *very* old, as old as life itself. For the other dramatic discovery of the late 1970s in molecular biology is that the genes of higher organisms are seldom 'recorded' in the chromosomes intact, but are scattered in fragmentary fashion along a stretch of DNA, broken up by chunks of DNA which seem at first sight to carry no message at all. It's as if this book started with a few pages of clear English, followed by an equally long, or longer, stretch of nonsense, then a few more pages of English picking up the story precisely where it left off, and so on. In order to read and use such a genetic message, the most fundamental ability of the cell, clearly going right back to the first appearance of cells, is the ability to cut out the chunks of nonsense and splice the pieces of coherent message together to make a working copy of the gene.

---

* Cambridge meeting reported by Roger Lewin in *Science*, volume 213, page 634, 1981. This discovery does not, of course, necessarily mean that human genes could get translated into plant chromosomes today, but that one of our ancestors may have 'given' this gene to an ancestral plant long ago. On the other hand, there is nothing in this discovery which says that transfer of genetic material from a human being to a cabbage, or the other way around, is impossible today.

# Split genes and spliced messengers

Split genes were first discovered by Pierre Chambon and his colleagues at the University of Strasbourg in 1977. They were investigating the way hens manufacture ovalbumin, the protein found in egg white; hens only make this protein when they are ready to lay eggs, so the genes responsible for manufacturing this protein must form an operon, a structural gene plus at least one regulatory gene which switches the structural gene on and off. Using the then new techniques of genetic recombination, Chambon's team set out to map the region of chromosome involved. At about the same time virologists at Cold Spring Harbor found that one particular viral gene they were investigating came in two parts, separated by a stretch of DNA that didn't seem to code for anything. News of this discovery arrived in Strasbourg just as Chabon's team were puzzling over their findings, that the messenger RNA produced by the region of chromosome responsible for manufacturing ovalbumin didn't seem to be big enough to carry all the information on that stretch of chromosome. The answer had to be that the mRNA was not, in fact, a direct copy of the entire stretch of chromosome, but only of the parts of the chromosome that carried the relevant genetic information. The intruding junk was being ignored.

We now know that the manufacture of mRNA happens in two stages. First, a stretch of chromosomal DNA is copied faithfully to produce a long stretch of RNA which includes copies of all the useless DNA (intervening sequences known as introns). Then, the cell's splicing and joining enzymes get to work, cutting out *precisely* the useless bits of RNA and very accurately joining the rest together to make a working molecule of mRNA that can be used by the cell to guide the manufacture of a particular protein. The precision of this process is absolutely vital, for a mistake here would make the mRNA useless. Somehow the cell recognises a particular nucleotide as the site for a cut, identifies the other end of the intron, perhaps five or six thousand base pairs away, equally precisely, loops the precursor RNA so that the pieces carrying genetic code (exons) are pulled together, and only then cuts out the loops of useless RNA and joins the pieces of exon together. Remember that the loss or addition of a single base pair would completely scramble the message in the genetic code, by shifting the reading frame in which the three-letter code words are translated. Such a mistake virtually never happens. And yet the introns *dominate* the length of the precursor RNA in very many cases. In the DNA carrying the ovalbumin gene itself, for example, there are seven introns; in another gene, for the protein conalbumin, there are 17 intruding sequences of nonsense, most of them longer than the 16 sequences of exon they surround; and in the gene for mouse beta-globin there is an intron 550 base pairs long, which is not only longer than any individual exon but longer than the final mRNA produced when all of the exons are joined together.

There are different schools of thought about why we should carry this load of genetic excess baggage around with us. Evolution by natural selection is a very efficient process, and there must be some advantage in having all this extra DNA, or species carrying the excess baggage would have lost out in the fight for survival to those that don't carry it. One school of thought has it that the splicing process itself is a means the cell uses to signal what is going on to other genes – that the genes, as it were, 'converse' among themselves. Another is that the introns play a regulatory role, not through any information they carry (they don't) but by their physical presence, rather like the way a fat protein molecule stuck on to the chromosome stops the production of beta-galactosidase. This is all new territory, and the only certainty is that we are in for more surprises as the research continues through the 1980s and beyond. But, just for fun, I'd like to tell you about the school of thought which seems to me to be the most promising at present. It originates with Nobel laureate Walter Gilbert, one of the leading actors throughout the history of molecular biology since the 1960s.

We saw in Chapter Three that the value of sex and

recombination in evolutionary terms is that they provide
new combinations of genetic material to be tested against
the cutting edge of natural selection. The presence of
introns within the genes indicates how easily genes themselves can be cut in half and recombined during this process, but it also holds out another possibility. A mutation
that caused the cell's machinery to make a slight mistake
during the processing of mRNA could easily shuffle the
exons of one gene into a new order on the final mRNA. As
always, most such rearrangements will be harmful – many
hereditary anemias, for example, seem to occur just
because the exons for the globin gene are put together in
the wrong order in the cells of sufferers. But, as with other
mutations, occasionally the new arrangement will produce
a more efficient protein. Gilbert estimates that this new
kind of 'recombination', the shuffling of the *parts* of a gene
into new combinations, could speed up the rate at which
proteins evolve a millionfold. There are good reasons why
rapid evolution at this level may be an advantage for
eukaryotes, which I will touch on in Chapter Ten. But the
fact that the presence of junk DNA allows shuffling of the
genome which gives an advantage today doesn't explain
how it arose in the first place. Gilbert has an answer to that,
too.

Think again about the emergence of the first living
things, the first self-replicating molecules of DNA (or
RNA; nobody can be quite sure which came first). The *first*
such molecules to form double helices in the primeval soup
must indeed have been stretches of nonsense DNA, coding
for nothing at all, just handfuls of bases stuck together at
random. What we think of as life began when the first short
stretches of DNA carrying a 'message' that influenced their
environment appeared. The first such messages – the first
genes – were simply stretches of nucleic acid that influenced their environment by encouraging the replication
process, perhaps by encouraging the formation of molecules that behaved like enzymes to break down passing
molecules into components that could more easily be used
to make more nucleic acid. It seems far more likely that
these first meaningful stretches of DNA should appear

within much longer stretches of nonsense than that the first
gene should appear, pristine and in splendid isolation, all
alone in the chemical soup. Whatever the conditions that
promoted the production of the first DNA molecules (and
nobody knows what they were), surely the earliest of those
molecules were, indeed, molecules of nonsense.

So, argues Gilbert, the first cells evolved around
stretches of DNA that consisted mostly of nonsense and
carried embedded within them the first useful genes. There
is no puzzle explaining how the nonsense DNA evolved,
because it got there first. What we really have to explain is
why one branch of evolution, the prokaryotes, got rid of
the nonsense and slimmed down into efficiently operating
single cells carrying the lightest possible load, while
another branch kept the extra DNA and developed into
the eukaryotes and ourselves. The question, thus phrased,
answers itself. There are two efficient ways for life on Earth
to follow. One sacrifices flexibility for faithful reproduction
and dogged sticking to the same lifestyle for thousands of
millions of years; the other favours flexibility, the opportu-
nism which enabled life to spread and occupy every available ecological niche. Neither can be said to be better than
the other; both are successful, as the presence of bacteria in
your gut and of your gut itself testify.

But brace yourself for some bad news. People are essentially protein, and protein molecules provide both the scaffolding for the body and the enzymes which ensure that the
body functions. Proteins are coded for by molecules of
DNA, housed on the chromosomes and passed on from
one generation to the next in sperm or egg cells. So far, so
good. But how much of that DNA is used to manufacture
the proteins necessary for the construction and functioning
of your body? Quite apart from the structural DNA that
codes for proteins, every cell has to have its quota of
structural DNA that codes for the cell's own workers, the
ribosomes, messenger RNA and tRNA among them.
Then, there are all the control genes responsible for turning the structural genes on and off, plus a great deal of
DNA that is not known to have any function, but may one
day be discovered to be an essential ingredient of life.

Whether or not all (or part) of the DNA found in introns is literally nonsense DNA, whether or not it is some sort of selfish parasitic lifeform, like an internal virus, hitching a free ride inside the cell, the important point is that all of this DNA which *doesn't* directly code for the proteins in our bodies is, by a long way, the majority of the DNA on our chromosomes, or those of a plant like maize. In your body, and mine, in fact, only one or two per cent of the DNA codes for protein.*

The possibility that now has to be taken seriously is that the DNA is part of an ecosystem within the cell, carrying on its business much as it did in the primeval ooze when life first evolved. The cell wall protects the interior from outside influences (except those the cell wants to let in), and maintains a stable environment for its contents for generation after generation. Within that environment, bits of DNA may be competing with one another, evolving still on the molecular level with no concern for what goes on in other cells outside; the genome itself is evolving not only under the pressure of Darwinian selection as the phenotype is tested in the world outside, but under the equally intense Darwinian pressure which ensures that only the 'fittest' *molecules* survive in that environment. Clearly, molecules which damage the cell or the whole organism excessively will bring about their own doom; but, equally clearly, as long as they don't disturb the machinery which ensures the healthy survival and reproduction of the cell and the organism it is part of, they can get a free ride through life. Some of this results in active evolution – genes that can jump from one chromosome to another, insert themselves and be copied indefinitely, achieving nothing except their own replication (which is, after all, the only criterion for success in evolutionary terms). Others are more passive. But only a tiny proportion of all the DNA is directly involved in creating the organism that is 'me' or 'you'. Maybe *we* are the excess baggage; just as a hen is only the egg's way of making another egg, a human being is

---

* See, for example, Roger Lewin, *Science*, volume 213, page 634.

just a cell's way of making more cells. What would you say is the most successful form of life on Earth? Humankind, which dominates the global environment so successfully? The bacteria which have stayed unchanged for three thousand million years or more? Or how about the ultimate parasite, the chunks of completely useless DNA, coding for nothing at all, that were the first self-replicating molecules to emerge from the ooze and that have been carried along from generation to generation, down through the eons, by all the busy activity of the 'useful' genetic material, evolving, adapting, breeding and passing on its genetic message to the next generation, while the nonsense gets passed on as well? Rather than being the pinnacle of creation, we might be regarded as extraneous appendages, bizarre offshoots from the process that really matters, the production of DNA. Our real role in life is to act as mobile homes, providing the equivalent of air conditioned luxury for the successful molecules inside our cells.

This is largely speculation at present. But even without understanding the exact role of all of the DNA in our cells, the existence of that DNA and the fact that it is all subject to the same sort of mutations – deletions, additions, translocations and inversions – provides molecular biologists with a remarkably accurate technique for monitoring evolution at work, and measuring the differences between species in terms of differences in their DNA. Indeed, the relevant techniques were being developed well before it was realised that a large proportion of the DNA in our chromosomes consists of non-coding introns. To use this tool, we don't need to know what any of the DNA does, only that it exists and that it has been around for a long time, gradually modified by mutation over an evolutionary timescale. There is no doubt that the introns in some eukaryotic genes have been there essentially unchanged for a very long time.

The globin genes that I have already mentioned provide a good example. There are two protein chains in human hemoglobin, the alpha and beta chains, which fold up to form similar three-dimensional components of the whole hemoglobin molecule. Each of these is also similar to the

structure of myoglobin, the protein in muscle cells that
holds on to oxygen. The similarities between the three
globins, and between the genes that code for them, suggest
that they have developed from a single ancestral type of
oxygen-carrying molecule, and that they split from this
common ancestor about a billion years ago. Each of the
three genes is made of three exons separated by two
introns, in the same parts of the gene in each case, and
those introns seem to have been preserved in place in each
gene for all of that billion-year history. Alec Jeffreys has
analysed in detail the DNA of the genes coding for the beta
chain, not just in human beings but in other members of the
primate family, our nearest living relatives, as well. He and
his colleagues found that the repeated sequences of DNA
making up the alleged 'junk' are very similar in the different species, and must have been conserved during evolution in just the same way as functional DNA – which
strongly suggests that they do have a function, even though
we don't yet understand it.*

These studies are very closely related to the work which
has provided the ultimate vindication of Darwin's ideas on
evolution and, specifically, on the origin of the human
species. The slight differences between the globin genes of
one primate species, such as ourselves, and another, such
as the orang-utan, are a result of millions of years of
diverging evolution since the two lines split off from a
common ancestor. The first thing the similarity in the
genetic material tells us is that there was, indeed, a
common ancestor. But that is only the beginning of the
story. Ever since the two species became distinct, mutations arising in one line have no longer been shared, by sex
and recombination, among members of the other branch of
the family. The longer the time since the split, the greater
the differences in the DNA, and in the proteins they code
for, between the two species. The remarkable discovery
stemming from dramatic, but still relatively unsung, work
in the 1960s and 1970s is that the build-up of mutational

* Work cited by Lewin, op. cit.

changes in each line is indeed *steady*, tiny differences adding up at a regular rate and ticking off the millennia since the two species diverged. By measuring the differences between the DNA of two living species today, it is now possible to say, with great confidence and very accurately, exactly when the two split from a common ancestor. And this molecular clock not only vindicated Darwin, it pointed the way to a dramatic revision of the timescale of human evolution, a revision now being confirmed by the fossil evidence uncovered by paleontologists in recent years.

# CHAPTER TEN

# FROM DARWIN TO DNA
## The molecular proof of human origins

When Charles Darwin published the *Origin* in 1859, it was only after years of careful thought, and even then under the pressure of Wallace's independent work. Darwin knew the impact his ideas were likely to have in a society still, to a large extent, dominated by the more reactionary kind of Church teaching, and even in 1859 he tried to steer around the worst of the likely storm of controversy by keeping humankind out of the story. All he said about our own evolution, almost at the end of his great work, was: 'In the distant future I see open fields for far more important researches . . . light will be thrown on the origin of man and his history.'* But the 'distant future' closed in on Darwin almost at once, with most of the furore about the *Origin* being directly concerned with the place of human beings on the evolutionary stage. In the 1860s, Huxley published an essay on the 'Evidence as to Man's Place in Nature', and in 1871 Darwin himself, the ground by now prepared, published *The Descent of Man*, in which he

* Pelican edition, page 458.

applied his theory of evolution by natural selection to the human species.

In the *Descent* Darwin sketched out the principles of natural selection, and pointed out the similarity of the human species to the living species of African apes – the gorillas and the chimpanzees. He said, 'we are naturally led to enquire, where was the birthplace of man?' And, pointing out that in all regions of the world today the living mammals are closely related to the extinct species of the same region, he concluded: 'it is therefore probable that Africa was formerly inhabited by extinct apes closely allied to the gorilla and chimpanzee; and as these two species are now man's nearest allies, it is somewhat more probable that our early progenitors lived on the African continent than elsewhere'.*

It would be difficult to provide a more succinct statement of modern opinion on human origins, except for changing 'somewhat more probable' to read 'virtually certain'. During the twentieth century, our understanding of human origins has been built up almost entirely through the study of those fossil ancestors to which Darwin alluded, and fossils clearly on the human lineage have been found in Africa and received wide publicity. In recent decades, the work of the Leakey family and the publicity given to Don Johanson's famous fossil 'Lucy' can have left few people – except those who, for whatever reasons, dismiss the idea of human evolution entirely – in any doubt that our ancestors evolved in Africa, and that we share a common lineage with the gorilla and chimpanzee.

But there is another side to this story, an attack on the puzzle of human origins which depends on the interpretation not of fragments of fossil bone but of the molecules in the blood and tissue of living species. This part of the tale is more recent (although it has respectable scientific predecessors going back almost to Darwin's time), and has received both less public attention and less than its due share of recognition from the fossil hunters themselves,

---

* Quotations from the second edition of *The Descent of Man*, page 155.

until very recently. But it is very close to the mainstream of the story I have been telling here, and it provides not only direct evidence of human evolution at work but also an accurate date for the evolutionary split when the human line began to diverge from those of the other African apes. The split can be dated and calibrated from comparisons of the DNA molecules in the cells of living people, living gorillas and living chimpanzees; and the molecular clock tells us that our common ancestor lived in Africa just five million years ago.

## Blood brothers

George Nuttall was born in San Francisco in 1862, right at the time, had he but known it, of the great debate about evolution stirred by the *Origin*. He grew up to do research in Germany, and become Professor of Biology at the University of Cambridge. In the early 1900s he provided the first direct evidence of the blood relationships between different species.

Nuttall used the then new discovery of the ability of the body to manufacture antibodies which protect it from attack by invaders. The first attack of a mild disease, such as chicken pox, makes the patient quite ill, but the body learns to identify the invaders that cause chicken pox, and to manufacture specific antibodies to destroy them. The next time the same invaders try to attack, the appropriate antibodies are produced immediately, and the illness never becomes noticeable. The body has become immune to chicken pox. But the antibodies that protect you against chicken pox are no protection against, say, influenza – indeed, the antibodies that protect you from one kind of 'flu virus may be little use against a different 'flu virus. Paul Ehrlich, the German who pioneered the use of antibiotics against disease, speculated in the 1890s that this specificity might make it possible to use the immune reaction to measure the blood relationships between species. But it was Nuttall who took the speculation up and made it practical reality.

He injected laboratory animals (the ones he used were rabbits) with samples of blood protein from a different species. The animal 'invaded' by the foreign material learned to manufacture antibodies against this specific invader, and the blood of the invaded animal provided a serum which reacted specifically with the blood proteins of the chosen invader. Although Nuttall knew nothing of the importance of DNA to evolution, and indeed little of antibodies themselves, he knew that this serum would react with other samples of the same blood to produce a dense precipitate in his test-tubes. But the precipitation reaction was far less strong when the blood of a different species of animal was treated with the specific serum. A serum prepared from rabbits 'invaded' with blood from a horse, say, reacted strongly with horse blood, but scarcely at all with cat blood. The strength of the reaction exactly followed the closeness of the similarity between different species, and Nuttall was quick to include human blood – his own – in the tests. After 16 000 tests involving species as diverse as fish and man, Nuttall reported to the London School of Tropical Medicine in 1901 that 'if we accept the degree of blood reaction as an index of blood relationship within the Anthropoidea, then we find that the Old World apes are more closely allied to man . . . exactly in accordance with the opinions expressed by Darwin'.*

I am not sure whether it is more surprising that this work was being carried out scarcely 40 years after the publication of the *Origin* and only 30 years after the publication of the *Descent*, or that it should then have lain dormant, with its potential failing to be exploited by evolutionary biologists, for more than half a century. The fact is, however, that the technique was not taken up until the late 1950s, when Morris Goodman, at Wayne State University in Detroit, applied the much more precise techniques of modern immunology to what were essentially the same experiments that Nuttall had pioneered. Goodman was able to measure the degree of blood relationships much more accurately than Nuttall, and he also had the benefit of 50 years of paleontology to draw on when comparing his measurements with the expected family tree. He found no surprises concerning the order in which different species had split off from the branch of evolution of which we are a part – the blood tests showed man and chimp to be very close relations, the gibbon to be a more distant cousin, the Old World monkeys to form a still more remote branch of the family, and the New World monkeys to be more distantly related yet. All this was very much what studies of the morphology of the living species suggested, and the fossils confirmed. The surprise came in a 1962 paper from Goodman, in which he tackled the problem of trying to sort out whether the chimp or the gorilla was more closely related to man. The short answer is that, from a variety of immunological tests, he found the three to be equally closely related.*

This came as a bombshell to biologists. Everyone accepted at the time – and a surprising number of biologists, let alone other people, still think to this day – that although the chimp and gorilla may be our nearest relations, 'obviously' they are much more closely related to each other than they are to us. That is what Goodman expected to find. But the molecules refused to conform to the expected view. They showed, and continue to show, that human, chimp and gorilla are mutually closely related, each one being equally close to the other two. You and I are as closely related to a chimpanzee as a gorilla is related to the chimpanzee. Even though the chimp and gorilla are hairy, inarticulate creatures that wear no clothes and live in the wilds of Africa (or in our zoos), while we are sophisticated, intelligent city dwellers who watch TV and eat frozen pizza (and capture chimps and gorillas for our amusement), the evolutionary difference between me and a gorilla is no more, and no less, than the evolutionary

* M. Goodman, 'Serological analysis of the systematics of recent hominoids', *Human Biology*, volume 35, page 377.

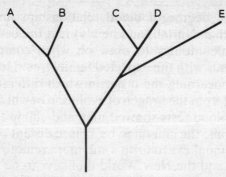

Figure 10.1 A hypothetical family tree, or cladogram. C and D are more closely related to each other than either is to E; but C, D and E are each equally distant cousins of A and B.

difference between a chimpanzee and that gorilla. The big question left unanswered by this immunological work was just when the ancestral lineage (represented in part, perhaps, by those fossils being avidly dug up at the time by the Leakeys and others) had diverged, in a three-way split, to produce the three lines that led to the three African apes, ourselves and our two hairy sibling species. The answer wasn't long coming, although it took a long time to get the paleontologists to begin to accept it. It came from a slightly different attack on the problem, one which brings us smack back on to the trail we followed throughout part two of this book, and which involves, in its early stages, a very familiar name.

# Molecular evolution

Linus Pauling's interest in hemoglobin had led to the discovery that a specific human ailment is produced by a change of a single amino acid in a protein chain, itself the result of a point mutation on a stretch of DNA coding for that protein. A single mistake in one codon of one gene produces the fatal disease, sickle cell anemia. Throughout the 1950s, and since, molecular biologists continued to study hemoglobin, for the same reasons that led Pauling's generation to its study. Hemoglobin is a large molecule, made of several subunits. 'Faults' in the manufacture of those components produce a variety of human ailments, so there will be direct practical benefits to medical science if the mechanisms producing those faults can be understood and corrected. In addition, because hemoglobin is produced in several slightly different forms even within an individual person – or an individual member of another species – it provides an opportunity to investigate how variations on the theme have arisen, and thereby to tackle the question of how variations on a molecular theme generally arise during the course of evolution. For all these reasons, Pauling's interest in the evolution of hemoglobin continued, and when he was joined at CalTech in 1959 by Emile Zuckerkandl (a native of Vienna who became a French citizen in 1938, and now works in France) he found a new collaborator with whom he could take his interest in hemoglobin a stage further.

Zuckerkandl took up Vernon Ingram's technique of 'fingerprinting' protein molecules using a mixture of electrophoresis and chromatography. With Pauling, he examined hemoglobins from several different species, identifying exactly how they differed from one another, where the chains of amino acids were identical and where, occasionally, there were different amino acids in the chains of one species compared with another. Goodman's technique used, more or less, whole blood and showed that we are very closely related to the chimp and gorilla. Zuckerkandl and Pauling looked at specific protein chains, and provided a direct measurement of the closeness of the relationship. They found that the difference between human hemoglobin and gorilla hemoglobin consists of a single substitution in one of the protein chains – an aspartine in the gorilla hemoglobin where there is a glutamine in human hemoglobin. This is, literally, the smallest possible difference – so small that it is virtually certain that somewhere on Earth there are human beings whose blood carries a mutated form of hemoglobin identical to normal gorilla hemoglobin. At this level, the differences between man and the other African apes appear to be no more significant than the variations among individual members of the human population.

It turns out that each of the mutations in variations on the hemoglobin theme has been produced by changes in a single base (not always the same base, of course), altering one letter in the genetic code. Statistical analysis of the pattern of these changes shows that there is no pattern – the mutations have occurred at random along the genes, leading to changes in the amino acids at random along the corresponding protein chains. And all of the evidence, including these crucial points, confirms that the extent of the evolutionary separation of two genes which have a common origin (a common 'ancestor gene') can indeed be measured in terms of the base changes occurring at random along a stretch of DNA. The variety of globin chains made it possible to establish the validity of this technique as an accurate indicator of the 'distance' between species – equivalent to the time since they split from a common ancestor.

But hemoglobin isn't the only molecule that can be studied in this way. Evolution proceeds through changes in the code carried by the DNA molecules of genes, but those changes in DNA are manifested in changes in the proteins that DNA codes for, and selection operates, effectively, at the protein level. The study of molecular evolution began as the study of protein differences between species and between individuals, and proteins still provide a wealth of information about evolution at the molecular level. From the mid 1960s onward, Walter Fitch and Emanuel Margoliash, of Northwestern University, near Chicago, have carried out detailed studies of the amino acid sequences of the protein cytochrome c in many species. Cytochromes are enzymes that are involved in the transport of energy-rich molecules around an organism, and like hemoglobin they exist in slightly different forms, doing more or less the same job, in a wide variety of species. Dog cytochrome c and horse cytochrome c, for example, differ by just ten amino acids in a chain 104 amino acids long. It is an interesting, and unresolved, question whether all of these changes are an evolutionary response to the different lifestyles of the two animals, or whether some are inconsequential mutations that have just happened by chance, so that a dog could get along just as well with horse cytochrome c as with its own. But, either way, these differences provide a measure of the differences between the two species, an indication of how far they have gone down separate evolutionary paths since they split off from some common ancestral stock.* Horse and dog are clearly less closely related than species that differ in their cytochrome c by only four or five amino acids. The cytochrome c family tree now provides an indication of the evolutionary relationships between chicken and penguin, tuna fish and moth, screwfly and turtle – and, of course, between human and ape. Always, the story is the same. Changes in the DNA coding for cytochrome c, like changes in the DNA coding for hemoglobin, have occurred at random, accumulating in the genome as time goes by. And human, chimpanzee and gorilla are practically identical, as closely related as it is possible for three species to be and to remain separate species.

Since the mid 1960s, the story has itself evolved along two lines. One takes the broad view, and looks far back in time to the origins of the genes that code for globin, the protein component of hemoglobin. The other, the line we are primarily interested in here, focuses down on that puzzling detail of the closeness of our own relationship with the African apes, and tries to fix that precise date for our split from the common ancestral stock.† Before we get

---

* And, as Thomas Jukes pointed out in 1966 in his excellent (though now slightly out of date) book *Molecules and Evolution* (Columbia University Press, page 192), reporting Margoliash's conclusions, the overall pattern of cytochrome c similarities and variations on the same basic theme provides 'striking evidence for the evolution of all the living species [so far studied] from a common ancestor'.

† The full story of the molecular clock and its impact on our understanding of human origins is told in *The Monkey Puzzle*, which I wrote with Jeremy Cherfas (Bodley Head, London; McGraw-Hill, New York; 1982). References to the main sources for this work can be found in that book.

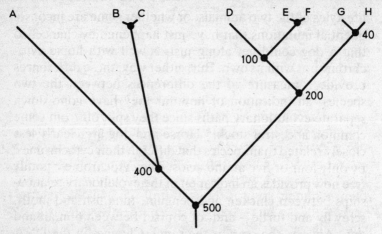

Figure 10.2 The degree of similarity between globins enables students of the molecular clock to calculate when the mutations that gave rise to the different lines occurred. The numbers are the times, in millions of years ago, that the different splits took place.

on to the timings of the molecular clock, it is perhaps worth a brief look at the latest news on the globin front, which takes on board what little we know, to date, about jumping genes and introns.

# The broad picture

Human hemoglobin, remember, contains two sets of each of two different protein chains, alpha and beta globin. But this isn't the whole story. There are actually five different 'beta like' globins, and three different 'alpha like' globins, which are manufactured by the body at different stages of development, from the early embryo to the fetus, early infancy and adulthood. It seems that, in response to the different oxygen requirements of the human organism at different stages during its life, an original globin theme has produced several different variations tailored to the specific needs of the body at different times. So how did they evolve?

The cluster of human genes coding for different forms of alpha globin are grouped together on one chromosome, arranged in the order in which they are 'switched on' during development. The beta cluster is similarly arranged, but on a different chromosome. And the similarities between alpha and beta globin genes suggest that they evolved from a common ancestral gene, beginning to diverge about 500 million years ago, at the very beginning of the evolution of the vertebrates. The alpha and beta genes both have two introns (regions of non-coding DNA), located in equivalent positions (one reason for suspecting their common origin); the layout of the genetic material suggests that there used to be a third intron, which got chopped out sometime during the course of evolution. That indication is particularly intriguing in view of the recent suggestion that plant leghemoglobin may have been transported into the plant genome as a hitch-hiker on a virus – for leghemoglobin has an extra, third, intron in exactly the place where vertebrate hemoglobin seems to have lost one. In that case, if there was any hitch-hiking involved, the plants must have acquired their new gene from some other source than vertebrates. The obvious candidates are insects, and studies of insect globin are now being carried out to find out if the extra intron is present there too.

The globin studies provide some of the clearest indications that introns are not, however, just junk. It is about 30 million years since the line of descent leading, on the one hand, to human and apes split from the line leading to the Old World monkeys, a date reliably confirmed by fossil evidence. Yet the beta cluster of genes in humans, gorillas and baboons still shows the presence of the same amount of 'extra' DNA in the same places, with very little change in the 'nonsense' code itself. The DNA may seem to be nonsense, and it certainly doesn't code for protein, but it must have some use, the argument runs, or mutations would scramble it much more rapidly than is seen to be the case.

Looking at more distantly related species, which is equivalent in some ways to looking back further in time, researchers such as Alec Jeffreys, of the University of

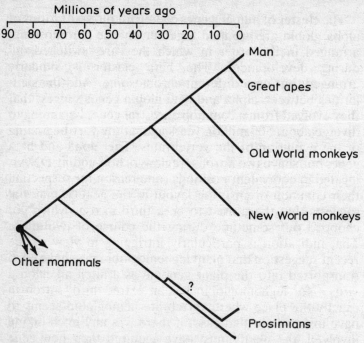

Figure 10.3 Focusing on just one of the globin gene clusters in the 'family tree' of globins in Figure 10.2 (it is beta globin, denoted on that figure by H) it is possible to calculate from the differences in their globins the dates when the human line split from the lines of the other primates. (Data for Figures 10.2 and 10.3 from A. J. Jeffreys, in *Evolution from molecules to men*, edited by D. S. Bendall, Cambridge University Press, 1983.)

Leicester, in England, have shown that the evolution of this whole cluster of genes has followed the simple pattern of a gene being duplicated so that the DNA carries an extra copy, with the new version of the gene then changing slightly through mutation until it produces a variation that is active at a different stage of development. Philip Leder, of Harvard Medical School, commented on these discoveries in *Science* in 1981:* 'The globin locus is constantly

* Volume 214, page 426.

pumping out copies of its genes and is bombarding the rest of the genome with them . . . what holds for globin must go for other genes too.' The complexity of human globin gene clusters is a direct feature of the evolutionary changes which make our gestation period so long, and which produce such helpless babies, still in a very underdeveloped state. In this sense, our globin genes are more 'evolved' than those of, say, the brown lemur, and the path of evolution from an ancestral form like that of the lemur to our own present form can be traced by looking at the genes of intermediate species, such as monkeys. And we can begin to see how evolution can proceed so efficiently, now that we know that the body doesn't rely on one copy of a working gene (which can't mutate without disrupting the regular working of the body) but has duplicate copies constantly being produced to act as the raw material of evolution. Spare copies of genes can mutate through all kinds of bizarre intermediate stages until they hit on a form that codes for a useful protein and gives the body they inhabit an edge which helps its selection. But I digress. Fascinating though this new work is, it doesn't tell us anything more about the timing of the split between man and the other African apes. It turns out that even Goodman's technique could provide the answer, although *proving* that the answer was correct took a little more effort.

# The molecular clock

In 1964 Vincent Sarich was a research student working in the Anthropology Department at the University of California, Berkeley. He participated in a series of seminars conducted by the professor of physical anthropology, Sherwood Washburn. Washburn was interested in the physical similarities between man and the other apes – the actual structure of the bones, the shape of the head, and so on. These physical similarities are, in spite of the obvious superficial differences, really very similar indeed. In a nutshell, the basic body pattern of man, chimpanzee and gorilla is the body pattern of a primate adapted to hanging from the branches and swinging under them (brachiation),* and the similarities between the three species are far greater than they ought to be if the human line split from the other apes 15 or 20 million years ago, as the conventional wisdom of the early 1960s had it, and has since been following a separate evolutionary path. Washburn wondered whether Goodman's technique of measuring the differences – or similarities – of blood proteins in the different species might not be used to provide some indication of how long ago, or how recently, this split actually occurred. And Sarich volunteered to read up all the available scientific papers on the subject (not that many in 1964) and report on them to the seminar group.

Sarich realised that if the mutations which had led to the differences in proteins in the blood and tissues of different species today had accumulated at a steady rate during evolutionary history then measurements of these differences could be used to indicate not only the evolutionary 'distance' between two species, but the time that had elapsed since they split from a common ancestor. The molecules could be used as a clock, with the accumulating mutations ticking away the millennia. The possibilities intrigued Allan Wilson, another member of the seminar group, who was just setting up a biochemical research team at Berkeley; Washburn suggested that Sarich take up the theme for his PhD topic, and Sarich duly joined Wilson's small group, beginning a collaboration that was to rewrite our understanding of the origins of humankind.†

The first key papers were published in 1967. Using biochemical techniques far more subtle even than those used

---

* We go into the details of all this in *The Monkey Puzzle*; among other things, the brachiating lifestyle, which involves hanging upright from branches, prepared our ancestors for the transition to upright *walking* when they descended to the ground.

† The full story of how that collaboration developed, and what it led to, can be found in the contributions from Sarich and J. E. Cronin, who joined the group later, in the volume *New Interpretations of Ape and Human Ancestry*, edited by Russel Ciochon and Robert Corruccini.

by Goodman, let alone by Nuttall, Sarich compared blood proteins from many different species, focusing particularly on the primates, our nearest relatives. He found that the difference between a human being and a gorilla or chimpanzee, in terms of the number of amino acid substitutions in the protein chains in their blood, is just one-sixth of the difference between a human being and an Old World monkey. Now, the date of the split between Old World monkeys and apes is one of the most reliable fossil dates, well determined at around 30 million years ago. If the mutations had indeed been building up at random and at a steady rate in the lineages studied over that time, there could be only one conclusion. The split between human and chimp (and gorilla) occurred only one-sixth as far back in the past as the split between all of the apes and the monkeys – that is, just five million years ago.

The very important requirement before this interpretation could be accepted was that the mutations which produced the changes in the amino acid sequences of the proteins must be occurring at random, at a steady rate. There was no problem about the randomness. Even in the mid-1960s it was clear that these differences are a result of random mutations of the DNA which codes for the proteins, and every study carried out since then has only added to the overwhelming weight of evidence that mutations do occur at random in the genome. But could the Berkeley researchers be sure that the *rate* at which these changes occurred stayed the same in different species even after they split off from a common ancestor? Suppose, for example, that the three-way split that produced human, gorilla and chimpanzee really had happened 15 million years ago, as the paleontologists would have accepted in 1967, but that since then evolution has, in a sense, proceeded only one-third as fast among the African apes as in closely related species, such as the Asian apes. Such a slowing down in the rate of evolution could produce the same apparent effect as a more recent split and a continuation of evolutionary change at the old rate.

This is a crucial point, because both in 1967 and in the years since many paleontologists have failed to understand

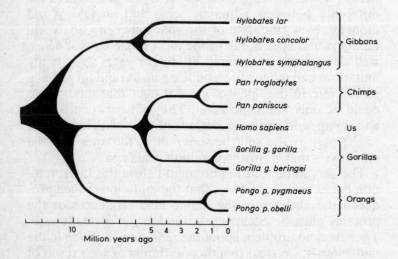

*Hylobates lar*

*Hylobates concolor*  } Gibbons

*Hylobates symphalangus*

*Pan troglodytes*  } Chimps

*Pan paniscus*

*Homo sapiens*  Us

*Gorilla g. gorilla*  } Gorillas

*Gorilla g. beringei*

*Pongo p. pygmaeus*  } Orangs

*Pongo p. obelli*

10   5  4  3  2  1  0
Million years ago

Figure 10.4 Many different versions of the molecular clock technique, using different molecules, combine to pinpoint our relationship to our closest kin in the animal world. It is less than five million years since the three-way split which produced man, chimps and gorillas from a common stock.

that Sarich and Wilson did not *assume* that the rate at which mutations accumulate – the rate of evolution – has stayed constant in the families of molecules they studied. They actually carried out experiments to *test* whether this is the case, and those experiments proved that the rate of mutation has stayed constant and that, therefore, these molecules can be used as a clock, reliably ticking off the timescale of human evolution.

How can the confusion have arisen in the minds of paleontologists? First, of course, there is the point that they didn't want to believe the new evidence – in 1967 they had what they thought to be a satisfactory picture of human origins, which pushed the split between man and the other African apes back to at least 15 million years ago. Two young biochemists with a new technique for dating the split between different species weren't going to be accepted with open arms, and the traditional timescale wasn't going to be rewritten overnight – not if the paleontologists could help it. Secondly, the way the proof of the molecular clock was published may have inadvertently delayed its full acceptance. As Sarich recalls, in 1966 and 1967 he and Wilson prepared three papers announcing their discovery. Two of these appeared in the prestigious and widely read journal *Science*, the second* dropping the bombshell of their five million year date for the split between ourselves and our nearest relatives. But in between those two papers Sarich and Wilson offered to the editors of *Science* another paper, one they regarded as vitally important, the one which presented the results of the tests which *showed* the rate of the molecular clock to be constant. In their wisdom, the editors of *Science* rejected this paper as 'not presenting anything unexpected'.† So this paper, presenting the critically important basis for the claim that the molecules could be read as a clock, actually appeared in the equally worthy but less widely read *Proceedings of the National Academy of Sciences*.‡ That article, says Sarich, 'has virtually never been referred to' in subsequent scientific papers criticising their molecular clock approach to evolution.

It may indeed be surprising that the molecular clock should tick at a steady rate – or rather, that each molecular clock should tick at its own steady rate, since each protein follows its own pattern of behaviour. One guess we might make, a natural assumption, would be that mutations accumulate more quickly in species that have a short generation time. Mice, on that picture, ought to 'evolve', at the

---

* Volume 158, page 1200.
† Quotation from Sarich, in *New Interpretations*, page 140.
‡ Volume 58, page 142.

molecular level, more rapidly than elephants do. That is
one guess, and to many people it seems a better guess than
the guess that the rate of accumulation of molecular mutations is the same, for a particular type of molecule, in all
species regardless of how quickly they breed. But the
important point is that those guesses can be tested, and that
the one that turns out to be correct, whether it seems
obvious or not, is that the clock ticks steadily.

The technique used to test these guesses is simple to
understand, although it involves a great deal of painstaking
work to carry through. Basically, it involves taking a well-
established evolutionary date, such as the split between the
apes and the Old World monkeys, and looking at the
changes that have accumulated in a chosen protein (the
first one Sarich and Wilson worked with was albumin) in
living representatives of as many species as possible that
are descended from the line which split at this well-known
time. In this particular case, the number of differences that
have accumulated in the albumins of all living apes, compared with Old World monkeys, is the same. The evolutionary distance from monkey to gibbon (which lives in
Asia) is the same as from monkey to chimpanzee (which
lives in Africa) or from monkey to man. Obviously, the
mutations are different in different species – the particular
amino acids that have mutated in the albumin of the orang-
utan, for example, are different from the ones that have
changed in the chimpanzee. But in all cases, including the
chimp and the orang, the *number* of mutations since the
split from the monkey line has been the same. The same
result is found when the comparison is made with New
World monkeys, or with carnivores, or with any species yet
tested.* The rate at which mutations accumulate in

---

* And, indeed, the technique has proved an invaluable tool in studying
the evolution of species as diverse as vampire bats, sea lions and the
panda, among our warm-blooded relations, and fish, reptiles and bacteria, among our more distant cousins. Wherever evolutionary origins of
living species are studied, Sarich and Wilson's molecular clock is a
standard and accepted tool. Only in the contentious area of human
origins, where the implications seem more personal, and to some people
more threatening, do a few traditionalists still hold out against the trend.

albumin in mammals depends only on the time that has elapsed, regardless of generation time or any other factor.

What is this rate? It varies from molecule to molecule, which makes it possible to check different molecular clocks against one another, and also to use the clock whose ticking is most suited to the species being studied. But, to take a couple of familiar examples, in cytochrome c change accumulates at a rate affecting one per cent of the protein chain every 20 million years, while hemoglobin shows a one per cent change every six million years, corresponding to a change in one amino acid every three and a half million years, on average. For what it is worth, even the fact that human and gorilla hemoglobin differ by only one amino acid fits the new timescale, although five million years is really too short an interval over which to apply the hemoglobin clock. In albumin, roughly every million years one amino acid, chosen at random along the chain, changes, thanks to a mutation in the DNA coding for albumin. Simply by counting the differences in amino acid composition of the protein chains, molecular biologists can count the time that has elapsed since two species shared a common ancestor. As Sarich has put it, 'what was a tentative best guess in 1967 had developed into a virtual certainty by 1970 – and remains so to this day'.*

All of the new data since then have confirmed the original date, give or take half a million years. Molecular anthropology, as it is now called, provides the best indication available of the date when our line first diverged from the lines leading to the gorilla and chimpanzee, and that date is just five million years ago. The icing on the cake of the new interpretation of human ancestry came when it proved possible to measure the differences not only in the protein coded for by DNA, but between the DNA molecules of different species themselves. And the DNA clock came up with exactly the same timescale for human evolution as all the other molecular clocks.

---

* Op. cit., page 141.

# DNA itself

One protein represents one gene at work, and one gene represents only about one two-millionth of all the information stored in human DNA. In addition, different proteins each have their own clock, mutations accumulating in each at its own rate. In spite of the great successes of the molecular clock approach, it would be much nicer if we could find a way to measure the accumulation of mutations in the DNA itself. Indeed, we can.

The technique goes back to 1960, when it was reported that DNA strands that had been broken apart by gentle heating would pair up again into double strands, bases pairing up in the usual manner, when allowed to cool. This, like the regularity of the molecular clock, may come as something of a surprise. The two strands in a double helix of DNA are only held together by the weak hydrogen bonds linking A with T and C with G, so it is no surprise that they can be parted by gentle heat. However gentle the heat, it is inevitable that the separate strands will get broken here and there in the process, and then the separate fragments of 'melted' DNA are free to coil up upon themselves as they wish. Making sense out of the resulting mess might seem a hopeless task. But when the brew is allowed to cool slowly again, the affinity between A and T and between G and C is so pronounced that the separated fragments of DNA line up with their partners once again, and re-forge the hydrogen bonds between each other. Two strands from a particular double helix are highly unlikely to find each other again, but if each one finds an exact replica of its former partner in the brew the effect is just as if the original helix has been renewed. The process by which the DNA fragments join up again is called annealing, and it is so effective that, if handled gently, the resulting DNA has much of its biological activity restored.

The annealing depends on the affinity between A and T and between C and G. Exactly complementary strands of DNA, the two halves of a whole double helix, clearly have a great affinity for each other because every A on one strand is matched by a T on the other, while every G on the other strand naturally lies alongside its preferred partner, the C with which it can make a strong hydrogen bond. The process is pure quantum physics at work – the atoms seek to form the arrangement which forms the lowest possible energy state, and thus form hydrogen bonds, perhaps the most quantum-mechanical of all chemical phenomena.

What would happen if the two strands of DNA trying to anneal were not perfectly matched? Obviously, some hydrogen bonds would still form where the bases did pair up, but in places where the 'wrong' bases lay opposite one another no bonds could form. The annealing would still occur, but less effectively, so that the double stranded molecules produced would not be held together so strongly. And that would mean that if they were heated once again they would be broken apart more easily than perfectly matched strands – such imperfectly annealed DNA would melt at a lower temperature than normal DNA. This is the key to the use of DNA as the ultimate molecular clock.

Two researchers who use the technique today are Jon Ahlquist and Charles Sibley, of Yale University. What they do is to take samples of DNA from each of two separate species – they might be man and chimp – and heat them to separate the DNA strands. The two sets of molten DNA are then mixed, and allowed to cool. As the brew cools, the single strands of DNA try to pair up, in line with the laws of quantum physics. Some will find their proper partners, and form tightly annealed helices; but in other cases a strand from one species will pair with a strand from the other species, forming a more loosely bound helix. When the solidified DNA is heated once again, these imperfectly combined strands will separate first, melting at a lower temperature than the rest. It is no mean feat to make the appropriate measurements and to identify just which DNA helices are melting when. But, glossing over the experimental details, two things are clear. The first is that ordinary DNA melts at a temperature of about 85°C. The second is that hybrid DNA formed by pairing one strand from one species with one strand from another melts at a lower temperature, and that each degree lower corresponds to one per cent difference in the string of bases along the DNA molecules. Two species that share 99 per cent of their DNA form hybrid DNA which melts at about 84°C; two species that have 98 per cent of their DNA bases in common (the same bases *in the same order* along the DNA) form hybrid DNA that melts at about 83°C. This test shows that human, chimpanzee and gorilla DNA are identical along at least 98 per cent of their length. The 'unique' features of humanity are contained in less than 2 per cent of our DNA.

Just as with the other molecular clocks, the DNA hybridisation test can be applied to species which are known, from reliable fossil evidence, to have split from a common ancestor at a certain date. The method is just the same as the one used by Sarich and Wilson for the albumin clock, and just as unambiguously informative. For example, all species where the fossil evidence unambiguously shows an evolutionary split 25 million years ago (the dog and the racoon provide a good example), the difference between the DNA of the two species amounts to about 12 per cent. The 2 per cent differences between human DNA and the DNA of either chimp or gorilla indicates a splitting time one-sixth of this, a little over four million years. Every test tells the same story. DNA changes do accumulate at a steady rate. The first measurements of this kind, applied to human, chimp and gorilla, indicated just the same three-way split, roughly five million years ago, that the protein clocks indicate. Now, the technique has been refined to the point where it is even more accurate for this particular task than the protein clocks. Ahlquist and Sibley carried out a study in the early 1980s which, they say, shows that the gorilla line split off first, about six million years ago, and that the split between human and chimpanzee lines occurred on the other branch of the family tree, four and a half million years ago. These figures are still entirely consistent with the protein clocks, but they shed just a little new light on human origins, and indicate, perhaps, just what the immediate ancestors of the human line looked like.

# Darwin vindicated

The molecular vindication of Darwin is twofold. First, Darwin's suggestion that the human line had its origins in Africa, and that the African apes are our closest living relations, has proved more accurate than even he can have guessed. The accumulating molecular data allow for no other conclusion than that the ancestor of our own line was also the ancestor of chimpanzee and gorilla, and there is a hint – as yet, no more than a hint – that chimp and human shared a common lineage for a brief time after the gorilla line split off. The growing weight of evidence – and a huge amount has been gathered since 1967 – cannot permit the human line to have diverged from the other African apes more than about five million years ago, unless the molecules of life, and the life molecule itself, have been behaving differently in these three species from those in all of the many hundreds of other species that have now been investigated.

The fact that such a tiny change in the DNA can account for the not inconsiderable difference between you and I and a gorilla indicates that the mutations which have produced human beings out of African ape stock must have involved control genes. Remember that most of our DNA is non-coding, and a good part of the 2 per cent difference between your DNA and that of a chimp must be accounted for by differences in the nonsense DNA. The proportion of *protein coding* DNA that differs in the two species may well be even less than 2 per cent. But it produces all the differences that make us human. The clue to how this can be comes from McClintock's work, and the more recent investigations of operators following on from the work of Jacob and Monod. In your body, only a small percentage of the DNA is active at any time; what matters is which bits are switched on, and when. By changing the way the operators do their work, the same basic blueprint that makes a chimpanzee can be modified to produce a human being. And we even have a good idea of what the modifications involve.

People are, in a very real sense, infant apes. We are born

immature, which makes us helpless as babies and requires adults to take care of their infants. This immaturity at birth is a crucial factor in allowing our brains to continue to develop and grow long past the point where the baby's head would be too big for its mother to give birth and live. Whereas most other mammals develop inside the womb to the point where they are fully formed and literally able to stand on their own feet and run with the herd immediately after birth, new-born human babies are in effect foetuses that still have some development to do, even though they are now outside the womb. Clearly, knowing what we now know about how genes are switched on and off, the crucial factor in the evolution of humans from the ape line has been a slowing down in development, a process called neoteny.

Instead of being born pre-programmed to run with the herd, we are born with the ability to learn and adapt to different circumstances, and the ability to develop a large brain with which to understand and control our environment. And the slowing down process doesn't stop there. A human being lives far longer than a chimpanzee but still retains features of a foetal ape, from the shape of our heads to the absence of hair on our bodies. From the point of view of a typical ape, humans have discovered, if not the secret of *eternal* youth, at least the secret of *prolonged* youth. And we owe it all to a very few genes that control the way in which other genes are switched on and off, and thereby the rate at which our bodies develop.

So what did the first proto-human – the first hominid – look like? The best evidence is that the chimpanzee is our closest relation, and that we both evolved from a common ancestor that was around four and a half million years ago. The oldest uniquely human ancestors were around in East Africa between three and four million years ago, and must have been rather like the oldest uniquely chimpanzee ancestors with which they shared the region. There are two species of chimpanzee alive today, *Pan troglodytes* and *Pan paniscus*, which diverged from a common line, the molecules tell us, between two and three million years ago. *Pan paniscus*, also known as the 'pygmy' chimp, has a smaller

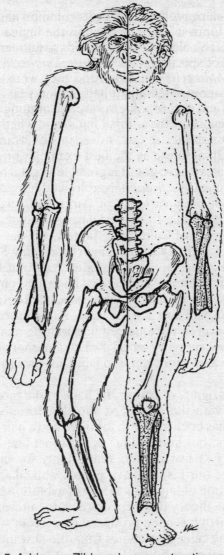

Figure 10.5 Adrienne Zihlman's reconstruction compares the structure of a modern pygmy chimp with that of *Australopithecus*, the ancestor of the human line. (Figure supplied by A. Zihlman and used with permission; see *New Interpretations of Ape and Human Ancestry*, ed. R. L. Ciochon and R. S. Corruccini, Plenum, New York, 1983, page 687.)

head than *Pan troglodytes*, hence its common name, but in terms of its limbs it is not as small as the image the name conjures up. The oldest fossil hominids so far identified are the remains of species labelled *Australopithecus afarensis* and *Australopithecus africanus*, and they were around in East Africa about three and a half million years ago. Paleontologists and anthropologists are still arguing about the exact significance of these fossil finds, and whether or not either *Australopithecus* was indeed a direct ancestor of ours. But, taking these as the best examples of our likely ancestors that we have got, it is possible to get a rough idea of what a typical *Australopithecus* looked like, combining features from both varieties of *Australopithecus* on the basis of the available fossil evidence. Adrienne Zihlman, of the University of California, Santa Cruz, has done just that, working with several colleagues including Sarich and Wilson, and comes up with a creature very much like the modern pygmy chimp. 'The earliest known hominids at 3.5 million years,' she concludes, 'may have been only one step away from a small ape like the living *Pan paniscus*.'* And really, if we weren't prejudiced into putting man, *Homo*, into a separate category on his own evolutionary branch, it would make a lot more sense to label our own species *Pan sapiens*.

Our real origins, of course, lie back in the more distant past, along with the origins of all other forms of life on Earth. Life has been on Earth for three and a half thousand million years, but human life, as a distinct line, has been around for just five million years. Only for one seven-hundredth of our history have people walked alone. The final vindication of Darwin's theory of evolution by natural selection, the theory that explains how the present variety of life on Earth has been produced by descent from those original living cells, also comes from the new understanding of molecular evolution.

---

* A. L. Zihlman and J. M. Lowenstein, in *New Interpretations*, page 691. This article, which begins on page 677, summarises the evidence for a chimp-like first hominid, and gives the references to the original research papers.

In the mid-1970s, the philosopher Karl Popper caused a flurry in the scientific dovecotes with a claim, made in his autobiographical *Unended Quest*,* that 'darwinism is not a testable scientific theory'. Of course, the remaining opponents of evolutionary ideas leapt upon this with delight, ignoring (or choosing not to appreciate) that Popper did not say that evolution does not occur, nor that Darwin was incorrect; Popper was simply splitting a philosophical hair about what constitutes a scientific theory.

The strict definition of a scientific theory requires that it should be capable of making predictions, and that the predictions can be tested – that is, that the theory must be inherently capable of being falsified. Darwin's theory of evolution by natural selection, said Popper, was no real theory because it simply explained the patterns of life we see about us, and in the fossil record, without making specific, testable predictions. In fact, Popper was mistaken, and his somewhat half-baked comments provoked responses from many scientists and philosophers of science, leading Popper to modify his position considerably, while most scientists agree that 'Darwinism' always was a respectable scientific theory, and still is today. It is indeed possible to make predictions, from Darwinian theory, about the changes that will take place in a laboratory population of, say, bacteria or fruit flies, subjected to different environmental conditions and therefore different selection processes. But the test I like best, which seems utterly to refute both Popper's original claim and the spurious arguments of those who invoke Popper as an ally in their attacks on evolutionary theorising, was carried out by three researchers in New Zealand, David Penny, L. R. Foulds and M. D. Hendy, and reported in *Nature* in 1982.†

Their argument runs like this. The sequences of amino acids in proteins provide evolutionary information, and

---

* Fontana, London, 1976. First published as 'Autobiography of Karl Popper', in *The Library of Living Philosophers*, edited by Paul Schilpp, Open Court Publishing Co., Illinois, 1974.
† Volume 297, page 197.

using this information it is possible to construct an evolutionary tree which shows when different species split off from the ancestral lines. From the information supplied by one type of protein – such as cytochrome c – it is possible, mathematically, to construct just one evolutionary tree that involves the minimum number of mutations, and therefore represents the most efficient way in which the observed pattern could have arisen. The fact that the data can be arranged in this way does not yet prove that evolution has occurred, only that an evolutionary tree does provide a possible explanation of the observed protein differences. But now comes the testable prediction: if evolution has occurred, then the minimal tree constructed by comparing cytochrome c from one species with cytochrome c from other species should be the *same* minimal tree that is found using data from other proteins but the same species. If all the protein trees were different, however, that would mean that the protein sequences do not contain evolutionary information, and would contradict the prediction made by evolutionary theory. The existence of a unique evolutionary tree is, therefore, a falsifiable hypothesis, and can be tested.

It should come as no surprise to anyone who has followed me this far to learn that when the New Zealanders looked at five different proteins from 11 different species they found all the evolutionary trees to be very similar. There are 34 459 425 different possible trees that could be drawn linking these 11 species, and although the five trees constructed by the New Zealand team were not quite identical to each other (the experimental techniques for measuring protein differences are not yet 100 per cent perfect), they are so very similar that, statistical tests show, there is only one chance in a hundred thousand that the similarity is a fluke. 'The different protein sequences give trees that are markedly similar, showing a relationship between them that is consistent with the theory of evolution.' Like all good theories, the theory of evolution is indeed testable; it has been tested, and so far it has passed every test.

Quantum physics provides an understanding of life at

the molecular level. It explains how proteins are put together and function in the cell, and how DNA helices coil and uncoil to manufacture messenger RNA or to duplicate themselves, passing on the hereditary message, the genetic code, from generation to generation. And now we see that those molecules whose properties have been so painstakingly investigated using the techniques of the physicist and quantum theory not only reveal the secrets of our own immediate ancestry, but also provide a direct test of the theory of evolution itself, a test Darwin's theory passes with flying colours. Where better to end my report on the links between quantum physics and life?

# Bibliography

These are the books which I read and consulted during the writing of the book you now hold. It is not a comprehensive bibliography. The literature on evolution is so extensive that this cannot be provided in a reasonable space, and the 'classical' works of the eighteenth and nineteenth centuries are under-represented, but all are referenced somewhere among the books which are cited. The *Encyclopaedia Britannica*, always an invaluable reference, is a particularly good quick guide to all of the historical material, for Hutton and Lyell's uniformitarianism through Darwin's own work and up to the discovery of the structure of DNA. The less accessible works, for those with a more specialised interest in the subject, are marked with an asterisk.

*D. S. Bendall, editor, *Evolution from Molecules to Men*, Cambridge University Press, 1983. Papers from the Darwin Centenary Conference held at Darwin College, Cambridge, in 1982, to mark the centenary of Darwin's death. Although primarily for the specialist, this includes interesting historical surveys which are of broad interest.

Peter Brent, *Charles Darwin*, Heinemann, London, 1981. The most up to date 'popular' biography, drawing on a wealth of material and giving the lie to the myth – still occasionally perpetuated – of Darwin as a playboy scientist who stumbled on the theory of evolution almost by chance.

*Elof Carlson, *The Gene: A Critical History*, W. B. Saunders, Philadelphia, 1966. A book for students of genetics, or historians of science, not for the casual reader. But very good at showing how each new idea developed.

G. M. Caroe, *William Henry Bragg*, Cambridge University Press, 1978. A biography of the older Bragg by his daughter (William Lawrence Bragg's sister). Only tangentially relevant to the story of the double helix, but includes an interesting chapter on the collaboration between father and son which established the nature of X-rays and set the scene for X-ray crystallography.

Jeremy Cherfas and John Gribbin, *The Redundant Male*, Bodley Head, London, and Pantheon, New York, 1984. A book about sex and evolution, in which we argue that the human male may have outlived his evolutionary usefulness. Along the way, however, the book goes into a great deal of detail about sex, recombination, and the value of the variety made possible by sex during evolutionary history. Fills out the picture sketched in chapter three of the present book.

*Russel Ciochon and Robert Corruccini, editors, *New Interpretations of Ape and Human Ancestry*, Plenum, New York, 1983. The proceedings from a landmark conference held in Italy in 1980, which marked the beginning of a rapprochement between the 'molecular clock' interpretation of human evolution and the traditionalists. A large (888 page) and mainly specialist tome, which includes a very clear account by Vincent Sarich of his pathfinding work and a concise overview, from J. E. Cronin, of the implications of the molecular clock for our understanding of human origins.

Helena Curtis, *Biology*, Worth, New York, second edition, 1975. Ostensibly a textbook, *Biology* is quite accessible for the interested layman, and weaves its tapestry of life using the unifying theme of evolution. A marvellous (but weighty – 1 000+ pages) overview of life on Earth.

Charles Darwin, *The Origin of Species by Means of Natural Selection*, Pelican, London, 1968. This edition of the classic work reproduces the text of the first edition of 1859, plus the 'Historical Sketch' and 'Glossary' which Darwin wrote later, and an Editor's Introduction from J. W. Burrow. It is easy to understand, written in clear language and should be read by everyone with an interest in evolution. Many other modern reprints are around, and most will do as well; try to find one based on Darwin's first edition, which is in many ways the strongest statement of his ideas.

Charles Darwin, *The Descent of Man*, John Murray, London, second edition, 1889. First published in 1871 (the second edition just happens to be the one that came to hand when I was working on this book), Darwin's second great work applies the theory of evolution by natural selection to our own species, and includes his conclusion that the origins of humankind lie in Africa, where our brother apes still live.

Charles Darwin, *The Voyage of Charles Darwin*, Ariel Books/BBC, London, 1978. Subtitled 'his autobiographical writings', this is a collection selected by Christopher Ralling to tell the story of the voyage of the *Beagle* and (very briefly) Darwin's later life, through his own words. A very good light read; thin on science.

Charles Darwin and Alfred Wallace, *Evolution by Natural Selection*, Cambridge University Press, 1958. The 'sketches' that Darwin wrote in 1842 and 1844, together with the joint paper read to the Linnean Society in 1858, collected in one volume with a foreword by Sir Gavin de Beer.

F. Darwin, editor, *The Foundations of the Origin of Species: Two Essays Written in 1842 and 1844 by Charles Darwin*, Cambridge University Press, 1909. Includes the *Sketch* of 1842 in which Darwin spelled out the theory of evolution by natural selection.

Francis Darwin, editor, *Life and Letters of Charles Darwin*, John Murray, London, 3 volumes, 1887. Includes an autobiographical chapter and T. H. Huxley's recollection of his reaction on first learning of Darwin's theory: 'how extremely stupid not to have thought of that'. A further collection was published in two volumes as *More Letters: Letters of Charles Darwin*, edited by F. Darwin and A. C. Seward, by the same publisher in 1903. Fascinating to dip into if you can find a copy in the library, and, of course, prime sources of information for historians of science.

*Paul Davies, *Quantum Mechanics*, Routledge & Kegan Paul, London, 1984. The best concise introduction to quantum mechanics for people with some background in science – at about first year undergraduate level. All the equations and a little of the physical and historical background make this an excellent student guide. Not for the casual reader, though; a mathematical tyro would be better off with John Polkinghorne's *The Quantum World* (see below), and the serious student could gain a lot by reading both books together.

Richard Dawkins, *The Selfish Gene*, Oxford University Press, 1976. One of the best explanations of how evolution works at the fundamental level, the level of genes. Dawkins also discusses the origin of life and the nature of mutations, explains how seemingly 'altruistic' behaviour of individuals can be a direct result of selfishness on the part of the genes the individual carries, and speculates that ideas – bits of information he calls 'memes' – may be the new replicators, evolving as ideas are copied and spread in the same way that genes evolve as they are copied and passed from generation to generation. Some of these theories have come under fire from his academic colleagues, but Dawkins' book is entertaining and insightful.

Gavin de Beer, *Charles Darwin*, Nelson, Edinburgh, 1963. The standard 'scientific' biography, still worth reading today, but rather thin on Darwin's early life, which gets more attention in Peter Brent's book. If possible, read them both to get a rounded view of the man *and* his work.

Loren Eisely, *Darwin and the Mysterious Mr. X*, Dent, London, 1979. A collection, published posthumously, of essays by a well-known naturalist/anthropologist about evolution and the principal pioneers of the subject. Eisely believed that Darwin had virtually stolen the idea of evolution by natural selection from a contemporary of his, Edward Blyth ('Mr. X'). His espousal of Blyth's cause certainly brought this 'unknown' father of evolution out of the shadows, but more careful research, stimulated in part by these claims, shows that Blyth did not, in fact, pre-empt Darwin (see the books by Lewis McKinney cited below). Nevertheless, Eisely provides a very good 'popular' introduction to the saga of how evolutionary ideas developed, provided that you take the championing of Blyth's cause with a pinch of salt.

*Ronald A. Fisher, *The Genetical Theory of Natural Selection*, Oxford University Press, 1930; revised edition, Dover, New York, 1958. A key book from a mathematician who was closely involved in developing the 'modern synthesis' of Darwinian natural selection and Mendelian genetics. Not an easy read, but a classic of its kind. Fisher's work provides the mathematical basis which proved that small, Darwinian variations in the individual members of a breeding population are completely adequate to account for evolutionary changes in the population through natural selection.

Antony Flew, *Malthus*, Pelican, London, 1970. Puts the famous *Essay on the Principle of Population* in perspective.

Edward Frankel, *DNA: The Ladder of Life*, McGraw-Hill, New York, second edition, 1979. A very clear, simple account of the chemistry of life, leaving out most of the historical complexity. Includes some excellent diagrams.

George Gamow, *Mr Tompkins in Paperback*, Cambridge

University Press, 1967. Entertaining popularisations of the ideas of relativity and quantum mechanics, from a physicist who was also involved in cracking the genetic code.

Wilma George, *Biologist Philosopher*, Abelard-Schuman, London, 1964. Subtitled 'A study of the life and writings of Alfred Russell Wallace', this is a readable account which emphasises Wallace's work, rather than the man, but doesn't go into so much scientific detail as the books by Lewis McKinney cited below.

John Gribbin, *Future Worlds*, Plenum, New York, 1981. Brings the Malthusian debate up to date, in the context of recent attempts to understand the immediate future of mankind.

John Gribbin, *In Search of Schrödinger's Cat*, Bantam, New York, and Wildwood House, London, 1984. The full story of quantum physics, including details of particle/wave duality and quantum uncertainty.

Jonathan Howard, *Darwin*, Oxford University Press, 1982. A brief introduction to the man and his work, published in the 'Past Masters' series of pocket book s.

Julian Huxley, *Evolution: The Modern Synthesis*, Allen & Unwin, London, 1942. Often reprinted and still readily available, Huxley's book provided the definitive statement of what has become the established view of evolution, the combination of Darwin's ideas on natural selection and Mendel's laws of genetics. Historically, the book is interesting because its publication as late as 1942 shows how long it took for a satisfactory, complete theory of evolution to become established. It is slightly less satisfactory to a modern reader as a summing up of evolutionary theory, because of the need Huxley felt to argue the case. But it is every bit as readable as Darwin's *Origin* – which, as an argument of a scientific case, it resembles somewhat – and it marks a convenient halfway house in scientific thinking between Darwin and today, a summing up of evolution just before the revolution in molecular biology.

Hugo Iltis, *Life of Mendel*, Allen & Unwin, London, 1932. The only book-length biography of Mendel, reprinted in 1966 and still fairly easily available. Very well researched, by an author who was a student in the early 1900s, at the time of Mendel's 'rediscovery', and pleasantly written, in a slightly old fashioned style. Iltis provides the best insight into Mendel's life anyone is likely to achieve, and also explains clearly the importance of his scientific work.

Horace Freeland Judson, *The Eighth Day of Creation*, Cape, London, 1979. A huge, baroque book, described by its author as 'an historical account of the chief discoveries of molecular biology'. In fact, it is more of a sourcebook for historians than a coherent history in its own right; it contains a wealth of information, but the reader has to dig deep to see how all the bits fit together. Wonderful to dip into, but infuriating if you attempt to read it from cover to cover. And best of all for its lengthy quotations from conversations between the author and key figures in the tale, including Bragg, Pauling, Watson, Delbrück and many more. A flawed diamond, but still one of my favourite books.

Evelyn Fox Keller, *A Feeling for the Organism*, W. H. Freeman, San Francisco, 1983. Subtitled 'The Life and Work of Barbara McClintock', this book was published just before McClintock was awarded her Nobel Prize. It is excellent as a biography, giving a feel for what it must have been like to be both a woman and a genius working in biology from the 1920s onwards. Not quite so good at explaining the biology, but still a superb book well worth reading by anyone interested in evolution.

John Kendrew, *The Thread of Life*, Bell & Sons, London, 1966. Based on a series of lectures on BBC TV in 1964, *The Thread of Life* provides a very readable, non-mathematic account of the revolution in biology from one of the major participants. Kendrew, with Max Perutz, determined the structure of the protein myoglobin, and received the Nobel Prize in Chemistry in 1962.

Albert L. Lehninger, *Principles of Biochemistry*, Worth, New York, 1982. One of the best, up to date texts designed for students taking an introductory undergraduate course in biochemistry, but actually very accessible to the interested general reader. The text is clear, the illustrations are excellent, and the only thing which might put you off is the sheer size of the book – just over a thousand pages. Well worth dipping into if you can find it in a library.

H. L. McKinney, editor, *Lamarck to Darwin: Contributions to Evolutionary Biology 1809–1859*, Coronado Press, Lawrence, Kansas, 1971. Contains reprints of papers from the era immediately before publication of the *Origin*. These include contributions from Wallace in 1855, Edward Blyth (the 'unknown' pioneer of evolutionary theory) in 1835, and the joint paper by Wallace and Darwin from 1858.

H. L. McKinney, *Wallace and Natural Selection*, Yale University Press, 1972. Probably the best account of how Wallace's work related to that of Darwin and his contemporaries. The discussion of the correspondence between Wallace, Darwin and Edward Blyth is especially interesting in the light of Loren Eisely's entertaining, but probably mistaken, contention that Darwin plagiarised Blyth's work. More scientifically detailed, but less readable, than Wilma George's book cited above.

Jacques Monod, *Chance and Necessity*, Collins, London, 1972. Based on a series of lectures given at Pomona College, California, in 1969. Nobel Laureat Monod gives a semi-philosophical, and very readable, overview of life and evolution.

Robert Olby, *The Path to the Double Helix*, Macmillan, London, 1974. The most authoritative history to date, much more accurate than Judson's *Eighth Day*, but less personal and less entertaining. Olby chooses, arbitrarily, to start his story in 1900, which rules out any proper discussion of Darwin and Mendel, and he ends, abruptly, with the identification of the double helix in 1953. Within those limitations, though, a good book in which to check out the facts of who did what, and said what to whom, where and when.

Colin Patterson, *Evolution*, Routledge & Kegan Paul, London, 1978. An up to date account, for a wide audience, by a zoologist based at the British Museum (Natural History) in London. Pulls no punches, but covers a lot of ground clearly in a short, well illustrated book.

*Linus Pauling, *The Nature of the Chemical Bond*, Cornell University Press, Ithaca, third edition, 1960. Originally published in 1939, this is Pauling's masterwork which establishes the basis of chemistry in terms of quantum physics, and describes the work which paved the way for the development of molecular biology. There is also a shortened version for students, which appeared from the same publishers in 1967 under the title *The Chemical Bond*. But both volumes are very much for scientific specialists, requiring at least undergraduate chemistry and physics, and are not immediately accessible to the general reader.

Linus Pauling and Peter Pauling, *Chemistry*, W. H. Freeman, San Francisco, 1975. Although it draws heavily on Linus Pauling's earlier books *General Chemistry* and *College Chemistry* (from the same publishers), this father/son collaboration brings the story right up to date (as of the mid 1970s) and provides the best introduction to chemistry that I know of. It requires scarcely any mathematics at all, is very clearly written and a good read, and it makes crystal clear the relationship between physics, chemistry and biology. Indeed, it would be difficult to imagine anyone ever producing a better book on the subject.

Max Perutz, *Proteins and Nucleic Acids*, Elsevier, Amsterdam, 1962. This book is based on the 1961 Weizmann Memorial Lectures, given by Perutz at the Weizmann Institute. It tells much the same story as Kendrew's *The Thread of Life* (see above), for an audience with more specialised knowledge of the subject. The two books complement each other nicely.

J. C. Polkinghorne, *The Quantum World*, Longman, London and New York, 1984. A short (100 pages), clear introduction to quantum physics, especially interesting because it is written by a man who gave up a professorship in mathematical physics at the University of Cambridge to become an Anglican priest. The ideal non-mathematical counterpart to Paul Davies' *Quantum Mechanics*.

Franklin Portugal and Jack Cohen, *A Century of DNA*, MIT Press, Cambridge, Massachusetts, 1977. A more coherent history than Judson's *Eighth Day*, although lacking some of the colour of that volume. Goes into great detail on the false dogma of the tetranucleotide hypothesis, and has very good, detailed references for anyone wishing to follow up the DNA story further.

Steve Prentis, *Biotechnology*, Orbis, London, 1984. Prentis is more concerned with the future applications of genetic engineering than the history of the subject, as indicated by his subtitle 'A new industrial revolution'. But he does provide a very clear quick overview of the chemistry of life and recombinant DNA techniques. Extremely well written and accessible; the best 'instant guide' to the topic.

Anne Sayre, *Rosalind Franklin & DNA*, W. W. Norton, New York, 1978 (original edition 1975). Written in response to Watson's *The Double Helix*, to set the record straight on behalf of Rosalind Franklin, Sayre's book doesn't claim to present a balanced view, but tries to redress the damage done to a dead scientist's image and reputation. It is a far better book than Watson's, and ought to be read by everyone who has heard Watson's side of the story.

Erwin Schrödinger, *What is Life?* and *Mind and Matter*, Cambridge University Press, 1967. Originally published as two separate books, in 1944 and 1958, respectively, and worth seeking out for the earlier contribution, in which Schrödinger explains the quantum mechanical basis of mutation, discusses the nature of the gene, and introduces the concept of the genetic code in its modern form. Written just before the realisation that DNA, not protein, carries the message of heredity, but still an excellent book, worth reading today in its own right, not just for its historical interest as the book which turned many physicists on to the possibilities of biological research.

Scientific American, *Organic Chemistry of Life*, W. H. Freeman, San Francisco, 1974. Readings from *Scientific American*, including a great deal about proteins and a good article on insulin.

G. Ledyard Stebbins, *Darwin to DNA, Molecules to Humanity*, W. H. Freeman, San Francisco, 1982. A very clearly written, comprehensive and up to date account of evolutionary ideas from one of the leading evolutionary biologists of recent decades. I know of no better treatment of the whole subject for a non-technical reader which takes full account of recent advances in molecular biology. Not a light read, but recommended if you want to get your teeth into the whole story.

Gunther Stent, editor, *The Double Helix*, Weidenfeld and Nicolson, London, 1981. A 'critical edition' of Watson's famous book, which includes the full original text, major contemporary reviews of the book from the 1960s (the original edition of *The Double Helix* appeared in 1968) by other people involved in the search for the double helix, some of the original scientific papers, and a commentary on the personal and publishing background to the story. You get the bonus of Watson's fast-moving version of the story,

reading like a good scientific thriller, together with an
authoritative presentation of the facts and context to the
work. Strongly recommended – but do read Anne Sayre's
*Rosalind Franklin & DNA* as well.

James Watson, *Molecular Biology of the Gene*, W. A.
Benjamin, Menlo Park, California, third edition, 1976.
Watson's excellent textbook, not to be confused with his
memoir *The Double Helix*. Really for serious students, but
well worth dipping into if you want to flesh out some of the
ideas only touched on in the present book.

James Watson, *The Double Helix*. See Gunther Stent.

James Watson, John Touze and David Kurtz, *Recombinant
DNA*, Scientific American Books, New York, 1983. A
comprehensive, clear and up to date (as of 1983) account of
how cells work and the techniques of genetic engineering.
The book is subtitled 'A short course', and aimed at serious
students; it is not always an easy read, not because the
authors lack skill but because they pack so much information into a short space. But it is well worth the effort if
you want to know why there has recently been so much fuss
about recombinant DNA techniques.

*Bruce Wheaton, *The Tiger and the Shark*, Cambridge
University Press, 1983. The story of the development of
the idea of wave-particle dualism (the 'tiger' and the
'shark' representing two creatures, each 'supreme in his
own element, but helpless in that of the other,' in a famous
quotation from J. J. Thomson). The story is fascinating,
but the asterisk here denotes not so much an impenetrable
wall of math as a very dense writing style which doesn't do
full justice to the tale. Worth dipping into, but not for the
fainthearted.

# Index

# READ MORE IN PENGUIN

## SCIENCE AND MATHEMATICS

### The Edge of Infinity   Paul Davies

Over the past decade, the evidence for black holes has greatly increased. In this updated edition, Paul Davies considers the latest research in this exciting and rapidly-developing field. At issue is the existence of boundaries not only to the physical universe, but maybe also to the very idea of what can be known and understood.

### The Newtonian Casino   Thomas A. Bass

'The story's appeal lies in its romantic obsessions ... Post-hippie computer freaks develop a system to beat the System, and take on Las Vegas to heroic and thrilling effect' – *The Times*

### Wonderful Life   Stephen Jay Gould

'He weaves together three extraordinary themes – one palaeontological, one human, one theoretical and historical – as he discusses the discovery of the Burgess Shale, with its amazing, wonderfully preserved fossils – a time-capsule of the early Cambrian seas' – *Mail on Sunday*

### The *New Scientist* Guide to Chaos   Edited by Nina Hall

In this collection of incisive reports, acknowledged experts such as Ian Stewart, Robert May and Benoit Mandelbrot draw on the latest research to explain the roots of chaos in modern mathematics and physics.

### Innumeracy   John Allen Paulos

'An engaging compilation of anecdotes and observations about those circumstances in which a very simple piece of mathematical insight can save an awful lot of futility' – *The Times Educational Supplement*

### Fractals   Hans Lauwerier

The extraordinary visual beauty of fractal images and their applications in chaos theory have made these endlessly repeating geometric figures widely familiar. This invaluable new book makes clear the basic mathematics of fractals; it will also teach people with computers how to make fractals themselves.

# READ MORE IN PENGUIN

In every corner of the world, on every subject under the sun, Penguin represents quality and variety – the very best in publishing today.

For complete information about books available from Penguin – including Puffins, Penguin Classics and Arkana – and how to order them, write to us at the appropriate address below. Please note that for copyright reasons the selection of books varies from country to country.

**In the United Kingdom**: Please write to *Dept. JC, Penguin Books Ltd, FREEPOST, West Drayton, Middlesex UB7 0BR.*

If you have any difficulty in obtaining a title, please send your order with the correct money, plus ten per cent for postage and packaging, to *PO Box No. 11, West Drayton, Middlesex UB7 0BR*

**In the United States**: Please write to *Consumer Sales, Penguin USA, P.O. Box 999, Dept. 17109, Bergenfield, New Jersey 07621-0120.* VISA and MasterCard holders call 1-800-253-6476 to order all Penguin titles

**In Canada**: Please write to *Penguin Books Canada Ltd, 10 Alcorn Avenue, Suite 300, Toronto, Ontario M4V 3B2*

**In Australia**: Please write to *Penguin Books Australia Ltd, P.O. Box 257, Ringwood, Victoria 3134*

**In New Zealand**: Please write to *Penguin Books (NZ) Ltd, Private Bag 102902, North Shore Mail Centre, Auckland 10*

**In India**: Please write to *Penguin Books India Pvt Ltd, 706 Eros Apartments, 56 Nehru Place, New Delhi 110 019*

**In the Netherlands**: Please write to *Penguin Books Netherlands bv, Postbus 3507, NL-1001 AH Amsterdam*

**In Germany**: Please write to *Penguin Books Deutschland GmbH, Metzlerstrasse 26, 60594 Frankfurt am Main*

**In Spain**: Please write to *Penguin Books S. A., Bravo Murillo 19, 1° B, 28015 Madrid*

**In Italy**: Please write to *Penguin Italia s.r.l., Via Felice Casati 20, I–20124 Milano*

**In France**: Please write to *Penguin France S. A., 17 rue Lejeune, F–31000 Toulouse*

**In Japan**: Please write to *Penguin Books Japan, Ishikiribashi Building, 2–5–4, Suido, Bunkyo-ku, Tokyo 112*

**In Greece**: Please write to *Penguin Hellas Ltd, Dimocritou 3, GR–106 71 Athens*

**In South Africa**: Please write to *Longman Penguin Southern Africa (Pty) Ltd, Private Bag X08, Bertsham 2013*

# READ MORE IN PENGUIN

**The Matter Myth**  Paul Davies and John Gribbin

Recent developments at the frontiers of science are challenging our views about ourselves and the nature of the cosmos as never before.

In this sweeping survey, acclaimed science writers Paul Davies and John Gribbin examine the revolutionary transformation that is currently overtaking scientific thinking. From the weird world of quantum physics and the theory of relativity to the latest ideas about the birth of the cosmos, they find evidence for a massive paradigm shift. Theories of black holes, cosmic strings, wormholes, solitons and chaos challenge common-sense concepts of space, time and matter and demand a radically new worldview. Here is a truly fascinating advance glimpse of twenty-first-century science.

**The Stuff of the Universe**  John Gribbin and Martin Rees
Dark Matter, Mankind and Anthropic Cosmology

In trying to make sense of our relationship with the cosmos, scientists have concluded that most of the universe is made up of so-called 'dark matter', the controlling factor in its dynamics, structure and eventual fate. In this illuminating account leading science writer John Gribbin and eminent physicist Martin Rees give us the most comprehensive and accessible treatment yet of the major theories and the latest advances in understanding the nature of dark matter, which lead on to the monumental question of why the universe is the way it is.

'The great question of Life, the Universe and Everything … a pleasure to read' – Tim Radford in the *Guardian*

**Stephen Hawking**  Michael White and John Gribbin

'Few scientists become legends in their own lifetime. Stephen Hawking is one. It's good to have this well-documented and immensely readable biography to remind us that the media-hyped "mute genius in the wheelchair" is in fact a sensitive, humorous, ambitious and occasionally wilful human being' – Paul Davies in *The Times Higher Educational Supplement*

# BY THE SAME AUTHOR

### In Search of the Edge of Time
Black Holes, White Holes, Wormholes

The phenomena now known as black holes were described as early as 1783 and dismissed as the fruit of idle speculation – invisible stars sounded just too implausible to be taken seriously. It was only with the development of radio astronomy, relativity theory and mathematical models of warped spacetime that their true significance became clear. Today, writes John Gribbin, 'virtually all astrophysicists regard black holes as a natural feature of our Universe'. Many believe they can function as tunnels leading to other times and other places and that they contain the key to the Big Bang; Stephen Hawking sees them as 'wormholes' linking mother and baby universes. Details of such theories are set out in this enthralling book, a guided tour through a still emerging cosmos of neutron and X-ray stars, white dwarfs, quasars and pulsars.

'Fascinating ... Gribbin's thought-provoking book is written in the smooth, easy style of professional science journalism' – *The New York Times Book Review*

### In the Beginning
The Birth of the Living Universe

Ripples in space collected by the COBE (Cosmic Background Explorer) satellite in 1992 clearly confirmed current ideas about the Big Bang. But why do matter and nature's fundamental forces seem specially designed to produce our kind of Universe? Some scientists see the hand of God; others call this a non-question; but Gribbin suggests a deeply satisfying new answer. Going far beyond the Gaia hypothesis that the Earth is a single living organism, he claims that galaxies may 'operate as supernova nurseries', that one universe can 'bud' from star-death and 'black hole bounce' into another, and that such 'offspring' are being steadily refined by evolution.

'Gribbin has combined his two main scientific passions – cosmology and evolution – to produce a fascinating book ... A superb overview of the very latest thinking of the universe' – *Sunday Times*

*also published:*

### Innervisions